# Math Made Clear: From the Basics to Calculus

Adriane Gründers

# Math Made Clear: From the Basics to Calculus

## 240 Key Topics Explained on a Single Page Each

 Springer

Adriane Gründers
Heidelberg, Baden-Württemberg, Germany

ISBN 978-3-662-73220-5          ISBN 978-3-662-73221-2   (eBook)
https://doi.org/10.1007/978-3-662-73221-2

# Why will this book help you?

**Who did I write it for?**

For everyone who doesn't like to read a lot, but wants or needs to understand math.

For everyone who likes things to be clear and well-organized.

For everyone who wants to find the essentials at a glance.

For everyone who thinks visually and likes to watch videos, but finds that videos aren't always ideal for learning math because you can't quickly look something up.

**What I want to tell you about math itself**

Something very important: In mathematics, everything fits together beautifully. But you only see that over time.

The calculation rules are not rules invented by people just so you can be tested on whether you follow them. Rather, they arise because they are the only way to calculate and do mathematics without contradictions.

Mathematics is logical in the sense that it is free of contradictions. But it is not logical in the sense that, when solving a problem, there is always a next step that can be logically derived. You need practice to see which next steps will help you move forward, and often there are several ways to reach the goal.

One more thing: Some people don't like math because they think you can't be as creative as you can with languages or art. That's not true— you might just need a little longer to get there. Even in writing or music, you first have to learn the language or the instrument, which also takes a few years.

**What makes this book special?**

I write in short sentences, just as I would explain it to you in person.

I try to explain everything using the simplest suitable example.

I go step by step—some things may seem obvious to you, but I promise that if you follow everything step by step, you will understand a lot in the end.

It's really important to think through and fully understand even the small steps. You wouldn't have learned to read so well if you hadn't first learned the individual letters.

Each page stands on its own: on the left, you'll find a short explanation; on the right, an example. Make the PDF page large enough so you can see the whole thing on your screen. If you're reading the printed book, it might help to cover the opposite page with a sheet of paper so you can focus on one page at a time.

I start right at the beginning, assuming no prior knowledge.

Even if you already know a bit, I recommend reading the simple pages at the start and checking whether you really understand everything, especially fractions and the binomial formulas. Mistakes often happen with the basics, and that doesn't have to be the case.

The individual pages are roughly built on each other and go from simple to more difficult. But I've tried to make sure you can also read the pages individually, as needed.

I've avoided cross-references between pages so you don't have to jump back and forth. Instead, I prefer to repeat some content.

Sometimes, especially in the chapter on functions, I also use terms and methods that come up later. You can skip them if they don't mean anything to you.

I am not trying to pull a rabbit out of a hat, but rather to explain everything from scratch, and also to include some things that you might find less often elsewhere, but which will help you to understand.

If you haven't understood something, it's often because no one has explained it to you in a way that you can understand. And I want to do that differently.

I am firmly convinced that you can understand the content of this book, and, building on that, much more mathematics if you wish.

Since I cannot know exactly what you need or want to know, it may be that you do not find everything you are looking for in these pages. However, based on the knowledge from this book, you will be able to easily continue your math journey by yourself.

Since I also do not know what is most important to you, I have deliberately not highlighted any formulae or rules. You will learn best if you use a highlighter yourself to mark the most important passages and add your own notes and further examples.

Sometimes I give an explanation directly below a mathematical expression or write out how to say it. It is helpful if you read the mathematical expressions aloud to yourself.

At the end of the book, I will give you a brief glimpse at complex numbers and the most beautiful formula in the world.

I wish you lots of fun and success with mathematics!

Adriane Gründers

PS: I would like to thank Iris Ruhmann, Stella Schmoll, and Bianca Alton for their professional and pleasant support from the publisher's side and for their many detailed suggestions. Sonja gave me valuable feedback, and Vanille Sperr and several other readers sent me numerous suggestions. Thank you all. For engaging in discussions to verify the English translation generated by artificial intelligence, I would like to express my gratitude to Zoë C, Alexandria Dobkowski, Celeen Rama Chandran, Christal Marr, Javier Ruiz, Patricia Sibeko, Kate Weddell, Terence Wood and Amira Benhouhou. I am always grateful for any suggestions regarding linguistic, didactic, and content improvements, as well as suggestions for additions which you can send to adrianegruenders@gmail.com.

# Contents


**Why will this book help you?** — v

**Basics and Calculations** — 1
  What are natural numbers? — 2
  How does addition work? — 3
  How does subtraction work? — 4
  Overview: Addition and subtraction — 5
  How does multiplication work? — 6
  In what order do you add and multiply? — 7
  How do you calculate with parentheses? — 8
  When can you simply leave out a parentheses? — 9
  How does the distributive law help you get rid of parentheses? — 10
  Why is minus times minus equal to plus? — 11
  Overview: Addition and multiplication — 12
  What do you need division for? — 13
  Why can't you divide by zero? — 14
  Why do you need fractions, i.e. rational numbers? — 15
  How do you expand and reduce fractions? — 16
  How do you add and subtract fractions? — 17
  How do you bring fractions to a common denominator? — 18
  How do you multiply and divide fractions? — 19
  Overview: Multiplication and division — 20
  Overview: Arithmetic of fractions — 21
  What are decimal numbers? — 22
  Why aren't fractions sufficient? — 23
  What are unknowns and variables? — 24
  How do you expand and factor expressions? — 25
  How can you rearrange equations? — 26

**Powers, Roots, Logarithms** — 27
  Why do you need powers? — 28
  What do negative exponents mean? — 29

How do you square parentheses? . . . . . . . . . . . . . . . . . . . 30
What are roots for and how do you calculate with them? . . . 31
What is the absolute value of a number? . . . . . . . . . . . . . 32
Why do you need logarithms? . . . . . . . . . . . . . . . . . . . . 33
What is a logarithm to a base other than 10? . . . . . . . . . . 34
What does it mean to write logarithm as log? . . . . . . . . . 35
How do you calculate with logarithms? . . . . . . . . . . . . . . 36
Why are there two inverse operations for exponentiation? . . 37
Overview: Exponents, roots, logarithms . . . . . . . . . . . . . 38

**Some Propositional Logic and Set Theory**                                    39
How do you calculate with "true" and "false"? . . . . . . . . . 40
What does "A implies B" mean? . . . . . . . . . . . . . . . . . . 41
What do "necessary" and "sufficient" mean? . . . . . . . . . 42
What does "if and only if" mean? . . . . . . . . . . . . . . . . . 43
What is a set? . . . . . . . . . . . . . . . . . . . . . . . . . . . . . . 44
How do you calculate with sets? . . . . . . . . . . . . . . . . . . 45
What are intervals? . . . . . . . . . . . . . . . . . . . . . . . . . . . 46
What are the most important types of mathematical proofs? . 47

**Solving Equations for Unknowns**                                              49
How to rearrange equations with equivalence rearrangements? 50
What are typical equivalence rearrangements? . . . . . . . . . 51
Where do you have to be careful when rearranging equations? 52
How do you recognize the type of an equation? . . . . . . . . 53
How do you solve a linear equation? . . . . . . . . . . . . . . . 54
How do you solve a quadratic equation? . . . . . . . . . . . . . 55
How does completing the square work in general? . . . . . . 56
How do you solve higher-degree equations? . . . . . . . . . . 57
How do you solve exponential equations? . . . . . . . . . . . 58
How to find the solution of an absolute value equation? . . . . 59
How do you determine the solution set of an inequality? . . . 60
How do you solve a system of linear equations? . . . . . . . . 61

**A Glimpse into Geometry**                                                     63
Why do the angles in a triangle sum up to 180° ? . . . . . . . . 64
What are similar triangles? . . . . . . . . . . . . . . . . . . . . . . 65
Where does the Pythagorean theorem help you? . . . . . . . 66
What determines the shape of a right triangle? . . . . . . . . . 67

How do you calculate the side lengths of a right triangle? . . . . 68
How is the sine defined and how do you calculate it? . . . . . . 69
How do you calculate $\sin(30°)$ and $\sin(60°)$? . . . . . . . . . . 70
What are cosine and tangent? . . . . . . . . . . . . . . . . . 71
Where do you find sine and cosine on the unit circle? . . . . . . 72
Where do the law of cosines and the law of sines help you? . . 73
What does area have to do with multiplication? . . . . . . . . 74
How can you remember the binomial formulas using areas? . 75
How do you calculate a volume? . . . . . . . . . . . . . . . . 76
What do exponents have to do with dimensions? . . . . . . . 77
What is $\pi$ and why do you need this number? . . . . . . . . . 78
What is radian measure and why is this unit practical? . . . . . 79
How do you calculate the area of a circle? . . . . . . . . . . . 80
Why do you need a coordinate system? . . . . . . . . . . . . . 81

**Solving Word and Application Problems**     **83**
What does math have to do with reality? . . . . . . . . . . . . 84
How do you solve word problems? . . . . . . . . . . . . . . . 85
How do you translate a word problem into mathematics? . . . 86
How do you solve the problem in mathematics? . . . . . . . . 87
How do you translate the solution back into reality? . . . . . . 88
What is the rule of three useful for? . . . . . . . . . . . . . . 89
How do percentage and interest calculations work? . . . . . . 90
How does compound interest work? . . . . . . . . . . . . . . 91
Why are units of measure important? . . . . . . . . . . . . . 92
How can you check formulas with units of measure? . . . . . . 93

**Functions**     **95**
What is a function? . . . . . . . . . . . . . . . . . . . . . . . 96
How can you represent a function? . . . . . . . . . . . . . . . 97
What does "injective" and "surjective" mean? . . . . . . . . . 98
What does "bijective" mean? . . . . . . . . . . . . . . . . . . 99
How do you get from a linear function to the line? . . . . . . . 100
What is the slope of a line and the slope triangle? . . . . . . . 101
How else can you get from a linear function to the line? . . . . 102
How do you get from a line to its equation? . . . . . . . . . . 103
How do you get the intercept form of a line? . . . . . . . . . . 104
What is the equation of the standard parabola? . . . . . . . . 105

How do you get from a quadratic function to the parabola? . . . 106
How do you determine whether a function is symmetric? . . . . 107
How do you shift functions? . . . . . . . . . . . . . . . . . . . . . . 108
How do you stretch functions in the $y$-direction? . . . . . . . . 109
How do you stretch functions in the $x$-direction? . . . . . . . . 110
How do you reflect functions over the line $y = x$? . . . . . . . 111
When are two lines perpendicular to each other? . . . . . . . 112
What do power functions look like? . . . . . . . . . . . . . . . . . 113
What is a polynomial? . . . . . . . . . . . . . . . . . . . . . . 114
How do you find the zeros of polynomials? . . . . . . . . . . . 115
What are multiple zeros? . . . . . . . . . . . . . . . . . . . . . . 116
What are rational functions? . . . . . . . . . . . . . . . . . . . . 117
How does polynomial division work? . . . . . . . . . . . . . . 118
How does polynomial division with remainder work? . . . . . 119
What do exponential functions look like? . . . . . . . . . . . . 120
What do logarithmic functions look like? . . . . . . . . . . . . 121
Overview: Injective, surjective, and bijective . . . . . . . . . . 122

**Sequences and Limits**     **123**
What are indices used for? . . . . . . . . . . . . . . . . . . . . . . 124
How do you calculate with summation signs? . . . . . . . . . 125
What is the binomial formula for higher powers? . . . . . . . 126
What is a sequence? . . . . . . . . . . . . . . . . . . . . . . . . . 127
How do you prove a statement for all natural numbers? . . . . 128
What does it mean for a sequence to converge to zero? . . . . 129
How do you check that a sequence converges to zero? . . . . 130
What does it mean for a sequence to converge? . . . . . . . . 131
What does it mean for a sequence to diverge? . . . . . . . . . 132
How do you determine whether a sequence converges? . . . 133
Why do monotone and bounded sequences converge? . . . . 134
What is a series and when does it converge? . . . . . . . . . . 135
How do you determine the value of a geometric series? . . . . 136
How can you determine the convergence of a series? . . . . . 137
What are limits of functions? . . . . . . . . . . . . . . . . . . . . 138
How do you calculate simple limits? . . . . . . . . . . . . . . . 139
What does it mean intuitively for a function to be continuous? 140
What is the rigorous definition of continuity? . . . . . . . . . . 141
What does it mean for a function to have a continuous extension? 142

How do you determine whether a function is continuous?   . .   143
How do you check a piecewise function for continuity?   . . . .   144

**Differential Calculus**   **145**
How do you determine the slope of a tangent? . . . . . . . . .   146
What does the derivative mean? . . . . . . . . . . . . . . . . .   147
When is a function differentiable? . . . . . . . . . . . . . . . .   148
How do you differentiate power functions? . . . . . . . . . . .   149
How do you differentiate sums of functions? . . . . . . . . . .   150
What is the slope of the exponential function at $x = 0$? . . . . .   151
What is special about the exponential function with $f'(0) = 1$?   152
How else can the $e^x$ function be written? . . . . . . . . . . . .   153
In what sense is the derivative a linearization? . . . . . . . . .   154
How do you use that the derivative is the linearization? . . . .   155
How do you differentiate a product? . . . . . . . . . . . . . . .   156
How do you differentiate a quotient? . . . . . . . . . . . . . .   157
What is a composite function? . . . . . . . . . . . . . . . . . .   158
How do you find the derivative of a composite function? . . .   159
How do you differentiate exponential functions? . . . . . . .   160
How do you find the derivative of the inverse function? . . . .   161
How do you determine the derivative of the inverse function?   162
Overview: Functions and their derivatives . . . . . . . . . . .   163
How do you find possible solutions for minima and maxima? .   164
How do you determine whether $f'(x) = 0$ gives min or max? .   165
What does the second derivative tell you? . . . . . . . . . . .   166
What are inflection points and how do you find them? . . . . .   167
Why is $(a_1 a_2 \dots a_n)^{1/n} \leq (a_1 + a_2 + \dots + a_n)/n$ ? (I) . . . . . .   168
Why is $(a_1 a_2 \dots a_n)^{1/n} \leq (a_1 + a_2 + \dots + a_n)/n$ ? (II) . . . . . .   169
Overview: Determining extrema and inflection points . . . . .   170
Overview: How are the graphs of f and $f'$ related? . . . . . . .   171
How do you solve optimization problems? (I) . . . . . . . . . .   172
How do you solve optimization problems? (II) . . . . . . . . . .   173
How do you do a curve sketching? . . . . . . . . . . . . . . . .   174
Overview: Curve sketching . . . . . . . . . . . . . . . . . . . .   175

**Integral Calculus**   **177**
How can you calculate areas using infinite sums? . . . . . . . .   178
What is a definite integral? . . . . . . . . . . . . . . . . . . . .   179

What is an indefinite integral? . . . . . . . . . . . . . . . . . . . . . 180
What is an antiderivative? . . . . . . . . . . . . . . . . . . . . 181
What does the fundamental theorem state? . . . . . . . . . . 182
Why does the fundamental theorem hold? . . . . . . . . . . . 183
How do you calculate antiderivatives? . . . . . . . . . . . . . 184
How can you calculate areas using antiderivatives? . . . . . . 185
How do you calculate integrals using integration by parts? (I) . 186
How do you calculate integrals using integration by parts? (II) 187
How do you calculate integrals using substitution? (I) . . . . . 188
How do you calculate integrals using substitution? (II) . . . . 189
How do you calculate the area of a circle? . . . . . . . . . . . 190
When is a function integrable? . . . . . . . . . . . . . . . . . . 191
When does an improper integral of a function exist? . . . . . . 192
Overview: Functions and their antiderivatives . . . . . . . . . 193

**Vector Algebra and Elementary Analytic Geometry**　　195
What is linear algebra and analytic geometry? . . . . . . . . . 196
What is a vector intuitively? (I) . . . . . . . . . . . . . . . . . . 197
What is a vector intuitively? (II) . . . . . . . . . . . . . . . . . . 198
How do you multiply vectors by a real number? . . . . . . . . 199
How do you add and subtract vectors? . . . . . . . . . . . . . 200
What is a linear combination of vectors? . . . . . . . . . . . . 201
When is a set of vectors linearly dependent? . . . . . . . . . 202
When is a set of vectors linearly independent? . . . . . . . . 203
What is the dot product? . . . . . . . . . . . . . . . . . . . . . 204
How do you determine the angle between two vectors? . . . 205
What does the dot product have to do with projection? . . . . 206
How can you represent a line in parametric form? . . . . . . . 207
How can you represent a plane in parametric form? . . . . . . 208
How can you represent a line in Hesse normal form? (I) . . . . 209
How can you represent a line in Hesse normal form? (II) . . . 210
How can you represent a plane in Hesse normal form? . . . . 211
How can you convert line equations into each other? (I) . . . 212
How can you convert line equations into each other? (II) . . . 213
Overview: Parametric and implicit equations . . . . . . . . . . 214

**Vectors, Matrices, and Systems of Linear Equations**  215

What is an (abstract) vector space? . . . . . . . . . . . . . . . . . 216
What is basis and dimension of a vector space? . . . . . . . . . 217
What is a linear map? . . . . . . . . . . . . . . . . . . . . . . . . 218
What do you need matrices for? . . . . . . . . . . . . . . . . . . . 219
How do you add matrices and multiply them by a scalar? . . . 220
How do you multiply matrices? . . . . . . . . . . . . . . . . . . . 221
How do you concretely multiply matrices? . . . . . . . . . . . . 222
What are important special cases of matrix multiplication? . . 223
What is the inverse matrix? . . . . . . . . . . . . . . . . . . . . . 224
What holds for matrix multiplication? . . . . . . . . . . . . . . . 225
What is a determinant? . . . . . . . . . . . . . . . . . . . . . . . 226
How can you understand $2 \times 2$ determinants? . . . . . . . . . 227
What properties does a determinant have? . . . . . . . . . . . 228
How can you understand $3 \times 3$ determinants? . . . . . . . . . 229
How can you calculate a $3 \times 3$ determinant? . . . . . . . . . . 230
How can you use determinants to calculate areas? . . . . . . . 231
What is the cross product? (I) . . . . . . . . . . . . . . . . . . . 232
What is the cross product? (II) . . . . . . . . . . . . . . . . . . . 233
What are the kernel and image of a linear map? . . . . . . . . 234
What does the rank of a matrix mean? . . . . . . . . . . . . . . 235
What does the dimension theorem state? . . . . . . . . . . . . 236
What does the dimension theorem say for endomorphisms? . 237
How do you calculate the solution of a $3 \times 3$ system? (I) . . . . 238
How do you calculate the solution of a $3 \times 3$ system? (II) . . . 239
How do you calculate the solution of a $3 \times 3$ system? (III) . . . 240
How do you represent a system of linear equations as a matrix? 241
What is the structure of the set of solutions of a linear system? 242
How do you calculate the inverse of a $3 \times 3$ matrix? . . . . . . 243
How to bring a linear system into row echelon form? . . . . . 244
What does the rank say about the solution set of a linear system? 245
What does the dimension theorem say for $A\mathbf{x} = \mathbf{0}$ and $\mathbf{x} \in \mathbb{R}^3$? 246
How do you calculate the intersection point of two lines? . . . 247
How do you calculate the point of line-plane-intersections? . . 248
How do you calculate the line of intersection between planes? 249
How to calculate the distance from a point to a line or a plane 250
How do you calculate the distance from a point to a line in 3D 251
How do you calculate the distance between two lines? . . . . . 252

Overview: Relative positions of points, lines, and planes  . . .  253
Overview: Products of vectors in different notations  . . . . . .  254

**Some Probability Theory**                                          **255**
What are events and how can you calculate with them? . . . .  256
What is probability? . . . . . . . . . . . . . . . . . . . . . . .  257
What does conditional probability mean? . . . . . . . . . . . .  258
How can you calculate a probability?  . . . . . . . . . . . . . .  259
What is a random variable and its distribution? . . . . . . . .  260
What are examples of discrete random variables? . . . . . . .  261
How do you add random variables? . . . . . . . . . . . . . . .  262
What is the expected value of a random variable? . . . . . . .  263
What is the variance of a random variable? . . . . . . . . . . .  264
How are the expected value and variance of $\overline{X}$ calculated?  . .  265
What do you need the binomial distribution for? . . . . . . . .  266
What does the binomial distribution look like? . . . . . . . . .  267
What is a continuous probability distribution?  . . . . . . . . .  268
What is the Gaussian normal distribution?  . . . . . . . . . . .  269
How can you understand the Gaussian normal distribution? (I)  270
How can you understand the Gaussian normal distribution? (II)  271

**What's next? – Outlook**                                           **273**
Are all infinite sets the same size? . . . . . . . . . . . . . . .  274
What are quantifiers useful for? . . . . . . . . . . . . . . . . .  275
What is algebra?  . . . . . . . . . . . . . . . . . . . . . . . . .  276
What do you need complex numbers for?  . . . . . . . . . . .  277
How do you calculate with complex numbers? . . . . . . . . .  278
How can you represent complex numbers on the plane?  . . .  279
How can you multiply complex numbers? . . . . . . . . . . . .  280
Why are complex numbers so important? . . . . . . . . . . . .  281
How can you represent functions by a power series?  . . . . .  282
How do you calculate Taylor series? . . . . . . . . . . . . . . .  283
How is the Taylor series related to the exponential function? .  284
How is $e^{ix}$ related to $\sin(x)$ and $\cos(x)$? . . . . . . . . . . .  285
How can you understand $e^{i\pi} + 1 = 0$ geometrically? . . . . . .  286
What is a differential equation? . . . . . . . . . . . . . . . . .  287
What is meant by "configuration space"? (I) . . . . . . . . . .  288
What is meant by "configuration space"? (II) . . . . . . . . . .  289

What is geometric algebra? . . . . . . . . . . . . . . . . . . . 290
What is the geometric product? . . . . . . . . . . . . . . . . . 291
What does the exterior product mean? . . . . . . . . . . . . . 292
How do you concretely calculate the geometric product? . . . 293

**Table of Overviews**     295

**If you are looking for terms or symbols, look here!**     297

# Basics and Calculations

This chapter is about addition, subtraction, multiplication, and division, i.e. the four basic arithmetic operations, and about how to calculate with whole numbers and fractions.

I also explain why you need parentheses and how you can get rid of them.

All of this is actually quite easy, but it's important to master it. Otherwise, you will make mistakes later on with simple things.

With the content covered here, you will already be able to solve simple word problems.


A. Gründers, *Math Made Clear: From the Basics to Calculus*, https://doi.org/10.1007/978-3-662-73221-2_1

# What are natural numbers?

The natural numbers are the numbers you use when counting. Sometimes, zero is also included among the natural numbers.

• • •...

1, 2, 3, ...

If you consider all natural numbers as a whole, this collection is called the set of natural numbers, abbreviated as $\mathbb{N}$. The special N indicates that it is a pre-defined set.

$$\mathbb{N} = \{1, 2, 3, \ldots\}$$

Sets are denoted on the left and right with curly brackets, and the elements of the set are listed in between.

$$M = \{\underbrace{2, 3, 5, 7, 11, 13, 17, 19}\}$$

↑      These are the elements.

set

If a certain element belongs to a set, this is written with a stylized small epsilon ($\in$, Greek e).

$$3 \in \mathbb{N}$$

3 is an element of the set of natural numbers.

If an element does not belong to a set, a crossed-out small epsilon is used.

$$4.37 \notin \mathbb{N}$$

The equality sign used above ("=") indicates that what is on the left is equal to what is on the right.

"=" means equality of what is on the left with what is on the right.
It originally comes from two parallel lines.

# How does addition work?

When you add two numbers , you simply count one after the other. If you first count up to 2 and then up to 3, this is the same as counting up to 5 in one go.

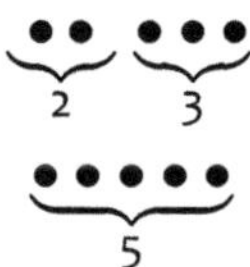

A plus sign "+" is used.

$$2 + 3 = 5$$

addend    sum

The numbers you add are called addends, the result is called the sum.

Since the order in which you count does not matter, the order does not matter in addition: you can swap the addends.

$$2 + 3 = 3 + 2$$

If you do not count any further (0 times), nothing changes.

$$5 + 0 = 5$$

It also does not matter whether you add two numbers and then a third, or the second and third and then the first.

$$(1 + 2) + 3 = 1 + (2 + 3) = 6$$

# How does subtraction work?

When you count backwards, you subtract. If you first count up to 5 and then count 2 backwards (or take away), it's the same as counting up to 3.

$$5 - 2 = 3$$

A minus sign "−" is used for subtraction.

If you count backwards more than you counted forwards, you get a negative number.

$$2 - 5 = -3$$

If you first count up to 2 and then take away 5, it's the same as if you count up to -3.

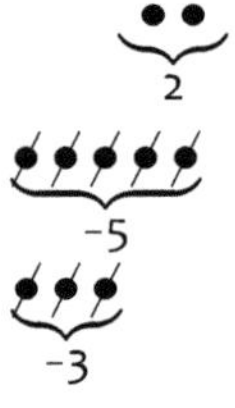

In subtraction the order matters.

$$5 - 2 \neq 2 - 5$$

A negative number is zero minus the corresponding positive number.

$$-3 = 0 - 3$$

If you add the negative numbers and zero to the set of natural numbers, you get the set of integers $\mathbb{Z}$.

$$\mathbb{Z} = \{\ldots, -3, -2, -1, 0, 1, 2, 3, \ldots\}$$

When you calculate with a negative number, you put parentheses around it. The parentheses can be omitted if the negative number is at the beginning.

$$7 + (-5)$$

$$-5 + 7$$

Subtraction is the inverse of addition.

$$5 - 3 = 2, \text{ since } 5 = 3 + 2$$

# Overview: Addition and subtraction

The letters in the following overviews are variables. You can substitute any numbers for them; in an equation, use the same numbers for the same letters.

$a, b, c$ are variables.

A term is a meaningful arrangement of numbers, variables, and operation symbols.

$(a + 5)^2$ is a term.
$+ {-})b$ is not a term.

You can omit parentheses when only adding.

$a + (b + c) = (a + b) + c = a + b + c$

You can switch the order in addition.

$a + b = b + a$

Adding zero does not change anything.

$a + 0 = 0 + a = a$

Subtraction of b is the same as adding –b.

$a - b = a + (-b)$

Subtracting the negative of a number is the same as adding that number.

$a - (-b) = a + b$

# How does multiplication work?

Multiplication with natural numbers is a shortcut for repeated addition.

$$3 \cdot 5 = \underbrace{5 + 5 + 5}_{\text{3 times the addend 5}}$$

You can imagine drawing 5 dots in a row, 3 times underneath each other, and just keep counting.

$$\left.\begin{array}{l} \bullet\ \bullet\ \bullet\ \bullet\ \bullet \\ \bullet\ \bullet\ \bullet\ \bullet\ \bullet \\ \bullet\ \bullet\ \bullet\ \bullet\ \bullet \end{array}\right\} \text{3 times}$$

If you multiply by 1, the number does not change.

$$1 \cdot 17 = 17$$

If you multiply by 0, the result is always 0.

$$0 \cdot 17 = 0$$

If you multiply by a negative number, the result is the negative of what you would get by multiplying the corresponding positive number.

$$(-2) \cdot 3 = -(2 \cdot 3) = -6$$

Therefore, a negative number is −1 times the corresponding positive number.

$$-3 = (-1) \cdot 3$$

This representation is often helpful.

In multiplication, you can switch the order: if you add 3 five times, it's the same as adding 5 three times.

$$3 \cdot 5 = 5 \cdot 3$$

You can see this by the fact that 3 rows of 5 dots (above) are the same as 5 columns of 3 dots (to the right).

$$\underbrace{\begin{array}{l} \bullet\ \bullet\ \bullet\ \bullet\ \bullet \\ \bullet\ \bullet\ \bullet\ \bullet\ \bullet \\ \bullet\ \bullet\ \bullet\ \bullet\ \bullet \end{array}}_{\text{5 times}}$$

Similarly, with a cuboid arrangement of dots, you can see that $3 \cdot (5 \cdot 7) = (3 \cdot 5) \cdot 7$.

# In what order do you add and multiply?

In addition the order never matters, even when adding negative numbers.

$$3 + 5 = 5 + 3$$
$$(-3) + 5 = 5 + (-3) = 5 - 3 = 2$$

Likewise the order does not matter in multiplication.

$$3 \cdot 5 = 5 \cdot 3 = 15$$
$$(-3) \cdot 5 = 5 \cdot (-3) = -15$$

You use parentheses to indicate the order in which different operations should be calculated.

To avoid the need for too many parentheses, it's convention, i.e. a generally accepted rule, that multiplication and division take precedence over addition and subtraction.

Multiplication ($\cdot$) and division ($\div$) take precedence over addition (+) and subtraction (−).

Here, you don't need any parentheses.

$$2 + 3 \cdot 5 = 2 + (3 \cdot 5) = 2 + 15 = 17$$

If you want it the other way around, that is, if you want to add first, you have to use parentheses.

$$(2 + 3) \cdot 5 = 5 \cdot 5 = 25$$

You can see that the result depends on the order in which you calculate.

$$2 + (3 \cdot 5) = 17 \neq 25 = (2 + 3) \cdot 5$$

The symbol "$\neq$" means "not equal", so there is no equality.

Depending on where you live, you may have encountered the order of operations being taught as PEMDAS, BODMAS or BEDMAS.

PEMDAS: **p**arentheses, **e**xponents, **m**ultiplication, **d**ivision, **a**ddition, **s**ubtraction

BODMAS: **b**rackets, **o**rders, **d**ivision, **m**ultiplication, **a**ddition, **s**ubtraction

BEDMAS: **b**rackets, **e**xponents, **m**ultiplication, **d**ivision, **a**ddition, **s**ubtraction

# How do you calculate with parentheses?

Parentheses indicate the order in which you calculate. An expression enclosed in parentheses is like a self-contained unit.

$$\ldots (5 + 6) \ldots$$

Calculate this before you continue.

If there are several parentheses, you calculate from the innermost parentheses outward.

$$2 \cdot (3 + 4 \cdot (5 + 6))$$

In the first step you calculate the innermost parentheses, i.e. $5 + 6 = 11$.

$$= 2 \cdot (3 + 4 \cdot 11)$$

Then you calculate the next innermost term, that is, $4 \cdot 11 = 44$.

$$= 2 \cdot (3 + 44)$$

And accordingly you continue from the inside out.

$$= 2 \cdot 47$$
$$= 94$$

Within the parentheses, or if there are none, you calculate from left to right.

For parentheses that are not nested, it doesn't matter in which order you calculate them.

$$(3 + 5) \cdot (4 - 2) =$$

$$= (3 + 5) \cdot (4 - 2)$$

# When can you simply leave out a parentheses?

You can delete the parentheses (or leave it out altogether) when there is only a single number (or a variable) left inside the parentheses.

$$2 \cdot (3 + 5) = 2 \cdot (8) = 2 \cdot 8 = 16$$

Note however, if the number is negative, you must take the minus sign into account (which means the number is multiplied by $-1$), so a plus in front becomes a minus.

$$4 + (5 - 8) = 4 + (-3) = 4 - 3 = 1$$

Furthermore, you can leave out the parentheses if there is a convention that defines the order, here multiplication and division ($\cdot$ or $\div$) precede addition or subtraction ($+$ or $-$).

$$2 + (3 \cdot 5) = 2 + 3 \cdot 5 = 2 + 15 = 17$$

Parentheses can be omitted due to multiplication-before-addition convention.

You can also leave out the parentheses if it doesn't matter in which order you calculate. This is only the case when you have several plus signs or multiplication dots in a row.

$$2 + (3 + 5) = (2 + 3) + 5 = 2 + 3 + 5 = 10$$
$$2 \cdot (3 \cdot 5) = (2 \cdot 3) \cdot 5 = 2 \cdot 3 \cdot 5 = 30$$

Caution: With minus signs (or division) you must not leave out the parentheses, as the result depends on the order.

$$2 - (3 + 5) = 2 - 8 = -6, \text{ but}$$
$$2 - 3 + 5 = -1 + 5 = 4, \text{ so}$$
$$2 - (3 + 5) \neq 2 - 3 + 5$$

Also, when multiplying expressions involving addition and parentheses, the parentheses must not be left out. The same applies when working with exponents (this will be explained later).

$$2 \cdot (3 + 5) \neq 2 \cdot 3 + 5$$
$$(2 + 3) \cdot (2 + 3) \neq 2 + 3 \cdot 2 + 3$$
$$(2 + 3)^2 \neq 2 + 3^2$$
$$(2 + 3)^2 \neq 2^2 + 3^2$$

# How does the distributive law help you get rid of parentheses?

You can always use the distributive law when there is a multiplication sign in front of a parenthesis and a plus or minus inside the parentheses.

$3 \cdot (2 + 3)$

The distributive law says that instead of adding first and then multiplying the sum, you can also multiply each term individually and then add the products.

$3 \cdot (2 + 3) = 3 \cdot 2 + 3 \cdot 3$
$3 \cdot 5 = 3 \cdot 2 + 3 \cdot 3$
$15 = 6 + 9$

You can easily see this using the visualization of multiplication.

So, if you want to get rid of parentheses around a sum, you use the distributive law.

$3 \cdot (100 + 3) = 3 \cdot 100 + 3 \cdot 3$
$= 300 + 9 = 309$

Sometimes it's also convenient to use the distributive law to create parentheses; this is called factoring out.

$7 \cdot 99 + 7 \cdot 1 = 7 \cdot (99 + 1)$
$= 7 \cdot 100 = 700$

A minus sign in front of a parenthesis is treated as multiplying what is inside the parentheses by −1.

$2 - (3 + 5) = 2 + (-1) \cdot (3 + 5)$
$= 2 + (-1) \cdot 3 + (-1) \cdot 5 = 2 - 3 - 5$

Similarly, it works when there is also a minus sign inside the parentheses, but you have to remember that minus times minus gives plus.

$2 - (3 - 5) = 2 + (-1) \cdot (3 - 5)$
$= 2 + (-1) \cdot 3 + (-1) \cdot (-5) = 2 - 3 + 5$

N.B. The distributive law applies to multiplication, but not in the same way to exponentiation.

$(1 + 1) \cdot 2 = 1 \cdot 2 + 1 \cdot 2$
$\underbrace{(1 + 1)^2}_{2^2=4} \neq \underbrace{1^2 + 1^2}_{1+1=2}$

# Why is minus times minus equal to plus?

The simplest explanation comes from applying the distributive law, i.e., the rule for how addition and multiplication work together.

We want to calculate the product of two negative numbers.

$$(-2) \cdot (-3)$$

We start with a simple equation that includes $-3$.

$$3 + (-3) = 0$$

To calculate the desired expression $(-2) \cdot (-3)$, we first have to create it. To do this, we multiply both sides by $-2$.

$$(-2) \cdot (3 + (-3)) = (-2) \cdot 0$$

On the left side we use the distributive law. On the right side, we note that any number multiplied by zero is again zero.

$$(-2) \cdot 3 + (-2) \cdot (-3) = 0$$

For the first term on the left, we use the fact that a negative number times a positive number gives a negative number.

$$-6 + (-2) \cdot (-3) = 0$$

Then we add six to both sides and have the result.

$$(-2) \cdot (-3) = +6$$

Therefore, minus times minus equals plus.

# Overview: Addition and multiplication

You can expand ...

$$a(b + c) = ab + ac$$
$$a(b - c) = ab - ac$$

... or factor (as above, just backwards).

$$ab + ac = a(b + c)$$
$$ab - ac = a(b - c)$$

With two parentheses you have to multiply every term in the first parentheses by every term in the second one.

$$(a + b)(c + d) =$$
$$a(c + d) + b(c + d) =$$
$$ac + ad + bc + bd$$

That means you multiply every term in the first sum by every term in the second sum.

|     | $c$   | $d$   |
| --- | ----- | ----- |
| $a$ | $ac$  | $ad$  |
| $b$ | $bc$  | $bd$  |

It's important to pay attention to the signs. When two minus signs are multiplied, the result is a plus sign. It's best to rewrite terms with a minus in front as terms with a plus in front, and the minus in parentheses.

$$a - b \cdot (d - e)$$
$$= a + (-b) \cdot d + (-b) \cdot (-e)$$
$$= a - b \cdot d + b \cdot e$$

Multiplication by $-1$ flips the sign.

$$(-1) \cdot a = (-1) \cdot (+a) = -a$$
$$(-1) \cdot (-a) = +a = a$$

Here I have already used variables instead of numbers.

# What do you need division for?

Division is the inverse of multiplication. If you want to know how many times 5 goes into 15, the answer is 3.

$$15 \div 5 = 3$$
since $3 \cdot 5 = 15$

For division, there are different notations: the division sign, the division slash, the fraction bar, or in some countries, the colon.

$$15 \div 5 = 3$$
$$15/5 = 3$$
$$\frac{15}{5} = 3$$
$$15 : 5 = 3$$

With the notations in a line, you must use parentheses when dividing by an expression, e.g. a sum.

$$15 \div (4 + 1) = 3$$
$$15/(4 + 1) = 3$$
$$15 \div (4 + 1) = 3$$

With the fraction bar, the bar itself acts as a kind of parentheses, grouping together whatever you are dividing by.

$$\frac{15}{4 + 1} = 3$$

From now on, we will always use the fraction bar, as it's the clearest.

Every number goes into itself exactly once, so dividing a number by itself gives one.

$$\frac{5}{5} = 1, \text{ since } 1 \cdot 5 = 5$$

In every number one goes into it as many times as the number indicates.

$$\frac{5}{1} = 5, \text{ since } 5 \cdot 1 = 5$$

# Why can't you divide by zero?

Division by 0 is not defined.

Therefore, do not divide by 0.

When you divide, check that the denominator is not equal to 0.

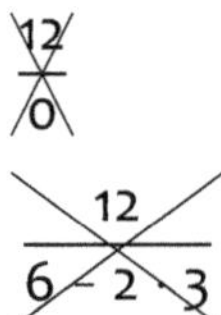

If you could divide a number by 0, the result would have to indicate what you need to multiply 0 by to get the original number.  But no matter what you multiply 0 by, you always get 0 and not the original number (e.g. 12).

$$\frac{12}{0} = ?$$

But: $0 \cdot ? = 0 \neq 12$

This consideration shows that at most you could divide 0 by 0.  But in that case, anything could result, since $a \cdot 0 = 0$.

$$\frac{0}{0} = ?$$

This cannot be calculated as is, only by using limits.

We will come back to this later: You can only define $\frac{0}{0}$ if both zeros have arisen from a common limiting process.

# Why do you need fractions, i.e. rational numbers?

If a natural number is an integer multiple of another natural number, the division again gives a natural number. We say the number is divisible by the other (without remainder).

$3 \cdot 5 = 15$

15 is a multiple of 5.

$$\frac{15}{5} = 3$$

If you divide a natural number by another that does not divide it, you do not get a natural number.

15 is not divisible by 4.
2 is not divisible by 3.
4 is not divisible by 6.

You then write the number as a fraction (or as a decimal, see below).

$$\frac{15}{4}, \frac{2}{3}, \frac{4}{6}$$

The upper number, the one being divided, is called the numerator.

In $\frac{3}{5}$, 3 is the numerator.

The lower number, the one you divide by, is called the denominator.

In $\frac{3}{5}$, 5 is the denominator.

You can write any fraction as the numerator times the fraction 1 over the denominator. This transformation can be helpful in both directions.

$$\frac{3}{5} = 3 \cdot \frac{1}{5}$$

If you divide 3 units into 5 parts, it's the same as dividing 1 unit into 5 parts and taking 3 of the parts (= fifths).

We call the set of all fractions rational numbers.

$\mathbb{Q}$ denotes the set of rational numbers, i.e. fractions.

By the way, a number that is only divisible by 1 and itself is called a prime number. The 1 is excluded.

$2, 3, 5, 7, 11, 13, 17, 19, 23, 29, 31, \ldots$
are prime numbers.

It can be proven that every natural number can be uniquely factored into prime numbers.

$1000 = 2 \cdot 2 \cdot 2 \cdot 5 \cdot 5 \cdot 5$
$2021 = 43 \cdot 47$
$123456789 = 3 \cdot 3 \cdot 3607 \cdot 3803$

# How do you expand and reduce fractions?

You expand a fraction by multiplying both numerator and denominator by the same number.

$$\frac{2}{3} = \frac{2 \cdot 2}{3 \cdot 2} = \frac{4}{6}$$

This is easy to understand:

If you divide 2 units into 3 parts, it's the same as dividing twice as many units into twice as many parts.

You can also imagine that you are multiplying the fraction by $1 = \frac{2}{2}$.

$$\frac{2}{3} = \frac{2}{3} \cdot 1 = \frac{2}{3} \cdot \frac{2}{2} = \frac{4}{6}$$

You reduce a fraction by dividing both the numerator and the denominator by the same number.

$$\frac{4}{6} = \frac{4/2}{6/2} = \frac{2}{3}$$

The easiest way is to factor the numerator and denominator so that the same factor appears in both; you can then simply cancel this factor, thus reducing the fraction.

$$\frac{4}{6} = \frac{2 \cdot 2}{2 \cdot 3} = \frac{\cancel{2} \cdot 2}{\cancel{2} \cdot 3} = \frac{2}{3}$$

If in the numerator there is only one number that you would cancel, then a 1 remains (since multiplication by 1 changes nothing and a fraction needs a numerator).

$$\frac{2}{4} = \frac{1 \cdot 2}{2 \cdot 2} = \frac{1 \cdot \cancel{2}}{2 \cdot \cancel{2}} = \frac{1}{2}$$

If in the denominator there is only one number that you would cancel, then at first a 1 remains, but you can omit it (since division by 1 changes nothing), and the fraction has become a natural number.

$$\frac{4}{2} = \frac{2 \cdot 2}{2 \cdot 1} = \frac{2 \cdot \cancel{2}}{\cancel{2} \cdot 1} = \frac{2}{1} = 2$$

# How do you add and subtract fractions?

Just as you can only add like terms, you can only add equal parts (same denominator).

2 sevenths + 3 sevenths = 5 sevenths

2 thirds and 3 fifths can't be added immediately.

Fractions with the same denominator can be added by factoring out 1/denominator.

$$\frac{2}{7} + \frac{3}{7} = 2 \cdot \frac{1}{7} + 3 \cdot \frac{1}{7} = (2 + 3) \cdot \frac{1}{7} = \frac{5}{7}$$

This means that you add the numerators and keep the denominator.

$$\frac{2}{7} + \frac{3}{7} = \frac{2 + 3}{7} = \frac{5}{7}$$

Fractions with different denominators must first be brought to a common denominator.

$$\frac{1}{50} + \frac{3}{100} = ?$$

To do this, you must multiply the numerator and denominator of one or both fractions by a number so that the denominators become equal.

$$\frac{1}{50} = \frac{2}{100}, \text{ in order for the}$$
two denominators to match

Then you can add the fractions by adding the numerators and keeping the (common) denominator.

$$\frac{1}{50} + \frac{3}{100} = \frac{2}{100} + \frac{3}{100} = \frac{2 + 3}{100} = \frac{5}{100}$$

You may then be able to reduce the fraction further.

$$\frac{5}{100} = \frac{1 \cdot 5}{20 \cdot 5} = \frac{1 \cdot \cancel{5}}{20 \cdot \cancel{5}} = \frac{1}{20}$$

Here is the example from the very beginning with the thirds and fifths. As explained, you have to expand both fractions so that the denominators become equal.

$$\frac{2}{3} + \frac{3}{5} = ?$$

$$\frac{2}{3} = \frac{10}{15} \quad \text{and} \quad \frac{3}{5} = \frac{9}{15}$$

$$\frac{2}{3} + \frac{3}{5} = \frac{10}{15} + \frac{9}{15} = \frac{19}{15}$$

# How do you bring fractions to a common denominator?

The common denominator is the least common multiple (LCM) of the denominators of the original fractions.

Multiples of 4: 4, 8, 12, 16, ...
Multiples of 6: 6, 12, 18, 24, ...
The least common multiple of 4 and 6 is 12.

When you add fractions and want to bring them to a common denominator, the most elegant way is to use the least common multiple.

$$\frac{3}{4} + \frac{5}{6} = \frac{3 \cdot 3}{4 \cdot 3} + \frac{5 \cdot 2}{6 \cdot 2}$$

$$= \frac{9}{12} + \frac{10}{12} = \frac{19}{12}$$

You can also simply multiply each fraction by the denominator of the other fraction.

$$\frac{3}{4} + \frac{5}{6} = \frac{3 \cdot 6}{4 \cdot 6} + \frac{5 \cdot 4}{6 \cdot 4}$$

$$= \frac{18}{24} + \frac{20}{24} = \frac{38}{24}$$

You may then need to reduce the fraction.

$$\frac{3}{4} + \frac{5}{6} = \frac{38}{24} = \frac{19}{12}$$

If you subtract fractions, it works the same way.

$$\frac{3}{4} - \frac{5}{6} = \frac{18}{24} - \frac{20}{24} = \frac{-2}{24} = -\frac{1}{12}$$

# How do you multiply and divide fractions?

You multiply a fraction by a whole number by multiplying the numerator by the number.

$$\frac{2}{3} \cdot 5 = \frac{2 \cdot 5}{3} = \frac{10}{3}$$

Dividing a fraction by a whole number is the same as multiplying the denominator by the number.

$$\frac{2}{3} \div 7 = \frac{2}{3} \cdot \frac{1}{7} = \frac{2}{3 \cdot 7} = \frac{2}{21}$$

You multiply by a fraction by doing both: multiplying the numerators and multiplying the denominators.

$$\frac{2}{3} \cdot \frac{5}{7} = \frac{2 \cdot 5}{3 \cdot 7} = \frac{10}{21}$$

You may be able to reduce, which is easiest to do before multiplying out the numerators and denominators.

$$\frac{2}{3} \cdot \frac{3}{4} = \frac{2 \cdot \cancel{3}}{\cancel{3} \cdot 4} = \frac{2}{4} = \frac{1}{2}$$

Multiplying by a whole number ($= \frac{\text{number}}{1}$) is a special case of this.

$$\frac{2}{3} \cdot 5 = \frac{2}{3} \cdot \frac{5}{1} = \frac{2 \cdot 5}{3 \cdot 1} = \frac{10}{3}$$

Dividing by a number is the same as multiplying by the reciprocal of the number ($= \frac{1}{\text{number}}$).

$$5 \div 3 = 5/3 = 5 \cdot 1/3 = 5 \cdot \frac{1}{3} = \frac{5}{3}$$

You get the reciprocal fraction by swapping the numerator and denominator of the fraction.

$$1 \div \frac{3}{4} = \frac{1}{\frac{3}{4}} = \frac{4}{3}$$

$$1 \div 3 = \frac{1}{\frac{3}{1}} = \frac{1}{3}$$

Dividing by a fraction is the same as multiplying by the reciprocal fraction.

$$\frac{2}{3} \div \frac{3}{4} = \frac{\frac{2}{3}}{\frac{3}{4}} = \frac{2}{3} \cdot \frac{4}{3} = \frac{2 \cdot 4}{3 \cdot 3} = \frac{8}{9}$$

# Overview: Multiplication and division

You can omit parentheses in expressions that only have multiplication dots.

$$a \cdot (b \cdot c) = (a \cdot b) \cdot c = a \cdot b \cdot c$$

You can swap the order in multiplication.

$$a \cdot b = b \cdot a$$

Multiplication by 1 does not change anything.

$$a \cdot 1 = 1 \cdot a = a$$

Multiplication by 0 makes everything 0.

$$a \cdot 0 = 0 \cdot a = 0$$

Division by b is the same as multiplication by $\frac{1}{b}$ = 1/b.

$$a/b = \frac{a}{b} = a \cdot \frac{1}{b} = a \cdot (1/b)$$

Division by 0 is not defined.

A fraction is a division. A fraction does not change if you multiply or divide both numerator and denominator by the same number (not 0).

$$\frac{a}{b} = \frac{a}{b} \cdot \frac{c}{c} = \frac{a \cdot c}{b \cdot c}$$

You multiply fractions by multiplying the numerators together and the denominators together.

$$\frac{a}{b} \cdot \frac{c}{d} = \frac{a \cdot c}{b \cdot d}$$

You divide fractions by multiplying by the reciprocal fraction (swap denominator and numerator).

$$\frac{a}{b} \div \frac{c}{d} = \frac{\frac{a}{b}}{\frac{c}{d}} = \frac{a}{b} \cdot \frac{d}{c} = \frac{a \cdot d}{b \cdot c} = \frac{ad}{bc}$$

Here I have used variables instead of numbers.

# Overview: Arithmetic of fractions

Fractions are multiplied by multiplying the numerators together and the denominators together.

$$\frac{a}{b} \cdot \frac{c}{d} = \frac{a \cdot c}{b \cdot d}$$

Fractions are divided by multiplying by the reciprocal fraction (swap numerator and denominator).

$$\frac{a}{b} \div \frac{c}{d} = \frac{\frac{a}{b}}{\frac{c}{d}} = \frac{a}{b} \cdot \frac{d}{c} = \frac{a \cdot d}{b \cdot c} = \frac{ad}{bc}$$

Fractions with the same denominator are added or subtracted by adding or subtracting the numerators and keeping the denominator.

$$\frac{a}{c} + \frac{b}{c} = \frac{a + b}{c}$$

$$\frac{a}{c} - \frac{b}{c} = \frac{a - b}{c}$$

Fractions with different denominators are added or subtracted by first making the denominators the same and then adding the numerators.

$$\frac{a}{b} + \frac{c}{d} = \frac{ad}{bd} + \frac{bc}{bd} = \frac{ad + bc}{bd}$$

$$\frac{a}{b} - \frac{c}{d} = \frac{ad}{bd} - \frac{bc}{bd} = \frac{ad - bc}{bd}$$

N.B. In contrast to multiplication, there is no distributive law for division, that is, no distributive law for denominators in fractions.

$$a \div (b + c) \neq a \div b + a \div c$$

$$\frac{a}{b + c} \neq \frac{a}{b} + \frac{a}{c}$$

For numerators in fractions, a distributive law holds, since division ($\div c$) corresponds to multiplication by the reciprocal ($\cdot \frac{1}{c}$).

$$(a + b) \div c = a \div c + b \div c$$

$$\frac{a + b}{c} = \frac{a}{c} + \frac{b}{c}$$

Here I have used variables instead of numbers.

# What are decimal numbers?

Decimal numbers are numbers with a decimal point.

$$234.567$$

To the left of the decimal point are the ones, tens, hundreds, thousands, etc. To the right of the decimal point are the tenths, hundredths, thousandths, etc.

$$234.0 = 234 = 2 \cdot 100 + 3 \cdot 10 + 4 \cdot 1$$

$$0.567 = 5 \cdot 0.1 + 6 \cdot 0.01 + 7 \cdot 0.001$$

There can be digits on both sides of the decimal point.

$$234.567 = 2 \cdot 100 + 3 \cdot 10 + 4 \cdot 1 + 5 \cdot 0.1 + 6 \cdot 0.01 + 7 \cdot 0.001$$

You can also write fractions as decimal numbers.

$$\tfrac{1}{2} = 0.5, \quad \tfrac{1}{4} = 0.25, \quad \tfrac{1}{20} = 0.05$$

Some fractions have an infinite number of digits after the decimal point. These repeat periodically. This is indicated by a bar over the repeating part.

$$\tfrac{1}{3} = 0.3333\ldots$$

$$\tfrac{1}{3} = 0.\overline{3} = 0.33\ldots$$

$$\tfrac{1}{7} = 0.\overline{142857} = 0.142857142857\ldots$$

If you want to write a terminating decimal as a fraction, multiply by a power of ten and possibly reduce it.

$$0.234 = \frac{0.234 \cdot 1000}{1000} = \frac{234}{1000} = \frac{117}{500}$$

If you want to write a repeating decimal as a fraction, you can do this by multiplying (depending on the length of the period) by 9 or 99 or 999, etc.

$$a = 0.234234234\ldots \text{ gives}$$
$$1000a = 234.234234234\ldots$$
$$-\ a = \phantom{234.}0.234234234\ldots$$
$$999a = 234,$$
$$\text{therefore } a = \tfrac{234}{999}$$

There are also decimal numbers whose digits do not repeat periodically or terminate; these cannot be written as fractions.

$$0.12345678910111213141516 17\ldots$$
$$\sqrt{2} = 1.414213562373095\ldots$$

The set of all decimal numbers are called the real numbers. They are all the points on the number line.

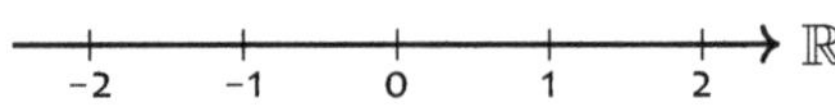

$\mathbb{R}$ denotes the set of real numbers, i.e. the set of decimal numbers.

# Why aren't fractions sufficient?

Decimal numbers that cannot be written as a fraction, i.e. not as a rational number, are called irrational numbers.

$$0.12345768910111213141516 17\ldots$$
$$\sqrt{2} = 1.4142135623 73095\ldots$$

You can quite easily understand why $\sqrt{2}$ is an irrational number, i.e. why it cannot be written as a fraction.

Suppose $\sqrt{2}$ could be written as a fraction, which we call $\frac{p}{q}$.

$$\sqrt{2} = \frac{p}{q}, \quad p, q \in \mathbb{N}$$

We decompose the numerator and denominator of the (assumed) fraction $\sqrt{2} = \frac{p}{q}$ into a power of 2 and an (odd) number.

$p = 2^r p'$, where $p'$ is odd.
$q = 2^s q'$, where $q'$ is odd.
$p$ contains the factor 2 $r$ times.
$q$ contains the factor 2 $s$ times.

We now square the original equation.

$$\frac{p}{q} = \sqrt{2} \Rightarrow \frac{p^2}{q^2} = 2$$
$$\Rightarrow p^2 = 2q^2$$

On the left side of the last equation, the factor 2 appears $2r$ times, so an even number of times, but on the right, the factor 2 appears a total of an odd number of times.

$$\underbrace{2^{2r}}_{\substack{2r \text{ times} \\ \text{the 2}}} \cdot \underbrace{p'^2}_{\text{odd}} = \underbrace{2 \cdot 2^{2s}}_{\substack{2s+1 \text{ times} \\ \text{the 2}}} \cdot \underbrace{q'^2}_{\text{odd}}$$

On the left, 2 appears an even number of times, on the right, an odd number of times.

But an even number cannot be equal to an odd number.

$$2r = 2s + 1$$
is a contradiction if $r, s \in \mathbb{N}$.

Therefore, $\sqrt{2}$ cannot be represented as the fraction of two integers, so it is an irrational number.

$\sqrt{2}$ is irrational.

# What are unknowns and variables?

What we have done so far does not only apply to the numbers we have used, but to any number.

$$2 + 3 = 3 + 2$$
$$17 + 23 = 23 + 17$$

We write a letter for any number. This is called a placeholder or variable. Each letter represents a fixed number within one calculation.

variables: $a, b, c, \ldots, x, y, z$

Calculation rules and procedures usually apply to many examples, actually an infinity of them. Thus they are formulated with variables.

$$a + b = b + a$$

We often use lowercase letters from the beginning of the alphabet for general variables.

$a, b$ etc. each stands for any number, always the same one within a given calculation.

When it comes to natural numbers, $n$ or $m$ is often used, and for indices we often use $i, j,$ or $k$.

$$n \in \mathbb{N}$$
$$i = 1, 2, 3, \ldots$$

When you are searching for a number, this initially unknown number, "the unknown", is often denoted by a letter at the end of the alphabet.

Which number, when multiplied by 3 and increased by 5, gives 12?

unknown number: $x$

You write the given requirement for the unknown number using $x$.

$$3 \cdot x + 5 = 12$$

When you multiply numbers with variables, the multiplication dot ($\cdot$) is usually omitted.

$$3x + 5 = 12$$

Likewise, when you multiply variables with each other.

$$ax + by = 1$$
instead of $a \cdot x + b \cdot y = 1$

# How do you expand and factor expressions?

The distributive law says that when you multiply a sum, you must multiply by each term.

$$a(b + c) = ab + ac$$

If the factor consists of a sum, you have to apply the distributive law twice.

$$(a + b)(c + d) =$$
$$a(c + d) + b(c + d) =$$
$$ac + ad + bc + bd$$

That means you multiply each term of the first sum by each term of the second sum.

| | $c$ | $d$ |
|---|---|---|
| $a$ | $ac$ | $ad$ |
| $b$ | $bc$ | $bd$ |

To determine whether you can factor out a variable, you need to check if it appears in every term.

$$ab + 5a = a(b + 5)$$

In both terms, $a$ appears so that you can factor out $a$.

It's best to mentally multiply it out again to see if it's correct.

$$a \cdot (b + 5) = a \cdot b + a \cdot 5 = ab + 5a$$

This is the same as before, so it's correct.

Often it's helpful to first choose one variable to factor out.

$$ab + 3b + 5a + 15 = a(b + 5) + 3b + 15$$
$$= a(b + 5) + 3(b + 5)$$
$$= (a + 3)(b + 5)$$

Sometimes it's not easy to see how you can factor something.

$$a^3 - b^3 = (a - b)(a^2 + ab + b^2)$$

If you reduce it to one variable (by dividing by $b^3$ and setting $x = a/b$), you get $x^3 - 1$. You can guess a root ($x = 1$) and then use polynomial division to divide by $x - 1$.

$$
\begin{array}{r}
(\quad x^3 \qquad\quad - 1) \div (x - 1) = x^2 + x + 1 \\
\underline{-x^3 + x^2} \qquad\qquad\qquad\qquad \\
x^2 \qquad\qquad\qquad\quad \\
\underline{-x^2 + x} \qquad\qquad\quad \\
x - 1 \qquad\quad \\
\underline{-x + 1} \quad \\
0
\end{array}
$$

# How can you rearrange equations?

If you have an equation and do the same thing to both sides, then both sides are still equal.

left side = right side
<u>do the same to both sides</u>
new left side = new right side

You can add a number to both sides.

$$5 - 3 = 2 \qquad | + 3$$
$$5 - 3 + 3 = 2 + 3$$
$$5 \cancel{-3} \cancel{+3} = 2 + 3$$
$$5 = 2 + 3$$

You can also subtract a number from both sides.

$$5 + 3 = 8 \qquad | - 3$$
$$5 + 3 - 3 = 8 - 3$$
$$5 \cancel{+3} \cancel{-3} = 8 - 3$$
$$5 = 8 - 3$$

Both ways are helpful when you want to solve an equation for $x$.

$$x + 3 = 8 \qquad | - 3$$
$$x + 3 - 3 = 8 - 3$$
$$x \cancel{+3} \cancel{-3} = 8 - 3$$
$$x = 8 - 3 = 5$$

You can also multiply both sides by a number.

$$\frac{15}{3} = 5 \qquad | \cdot 3$$
$$3 \cdot \frac{15}{3} = 3 \cdot 5$$
$$15 = 3 \cdot 5$$

If you have a fraction on both sides of the equation, you can multiply by both denominators. It's called "cross-multiplication".

$$\frac{3}{5} = \frac{12}{20}$$
$$3 \cdot 20 = 12 \cdot 5$$

If you multiply by 0 you can turn a false equation into a true one. Thus, multiplying by 0 is not an equivalent rearrangement.

3 = 5 is a false equation
$$0 \cdot 3 = 0 \cdot 5$$
0 = 0 is a true equation

You can divide both sides by a number. This number must not be 0.

$$3x = 12 \qquad | \div 3$$
$$\frac{3x}{3} = \frac{12}{3}$$
$$x = 4$$

# Powers, Roots, Logarithms

After understanding the four basic arithmetic operations, we now add exponentiation, i.e. raising to a power, as in 3 to the $5^{\text{th}}$ power, $3^5$.

Just as repeated addition results in multiplication, repeated multiplication results in a power.

There are two inverse operations: taking roots and taking logarithms. I will explain this all in detail shortly.

Powers provide important functions for many applications: the power function, when the variable is in the base, and the exponential function, when the variable is in the exponent.

© The Author(s), under exclusive license to Springer-Verlag GmbH, DE, part of Springer Nature 2026
A. Gründers, *Math Made Clear: From the Basics to Calculus*,
https://doi.org/10.1007/978-3-662-73221-2_2

# Why do you need powers?

Exponentiation with natural numbers is a shortcut for repeated multiplication.

$$7^3 = \underbrace{7 \cdot 7 \cdot 7}_{\text{3 times the factor 7}}$$

The number, which is written below and is multiplied with itself, is called the base. The number written above, which indicates how many times you multiply, is called the exponent.

$$7^3 \leftarrow \text{exponent}$$
$$\uparrow$$
$$\text{base}$$

If you first multiply 7 with itself 3 times and then again 2 times, it's the same as multiplying 7 with itself 5 times.

$$\underbrace{7 \cdot 7 \cdot 7}_{7^3} \cdot \underbrace{7 \cdot 7}_{7^2} = \underbrace{7 \cdot 7 \cdot 7 \cdot 7 \cdot 7}_{7^5}$$

$$7^3 \cdot 7^2 = 7^5$$

This also holds in general.

$$a^n \cdot a^m = a^{n+m}$$

If you multiply 7 with itself 0 times additionally, nothing changes, it's the same as multiplying by 1. Thus, $7^0 = 1$.

$$7^3 \cdot 7^0 = 7^{3+0} = 7^3 \cdot 1$$
$$7^0 = 1$$

This holds in general: Any number to the power of 0 is equal to 1 (as long as the number is not zero).

$$a^0 = 1 \text{ for } a \neq 0$$

If you first multiply 7 with itself 5 times and then divide by 7 twice, it's the same as multiplying 7 with itself only 3 times.

$$\frac{7^5}{7^2} = \frac{7 \cdot 7 \cdot 7 \cdot \cancel{7} \cdot \cancel{7}}{\cancel{7} \cdot \cancel{7}} = 7^3$$

This holds in general.

$$\frac{a^n}{a^m} = a^{n-m}$$

It's also valid for $n = m$.

$$\frac{a^n}{a^n} = a^{n-n} = a^0 = 1$$

# What do negative exponents mean?

If there is a negative number in the exponent, it's the same as taking the reciprocal of the power with a positive exponent.

$$a^{-n} = a^{0-n} = \frac{a^0}{a^n} = \frac{1}{a^n}$$

In particular, $a^{-1}$ is just another way of writing $\frac{1}{a}$.

$$a^{-1} = \frac{1}{a}$$

If you raise a power to another power, you have to multiply the exponents.

$$(a^b)^c = a^{bc}$$

You can understand this if you consider that exponentiation is a shortcut for repeated multiplication.

$$(2^3)^2 = \underbrace{(2 \cdot 2 \cdot 2)}_{2^3} \cdot \underbrace{(2 \cdot 2 \cdot 2)}_{2^3}$$
$$\underbrace{\phantom{(2 \cdot 2 \cdot 2) \cdot (2 \cdot 2 \cdot 2)}}_{(2^3)^2}$$

You must pay attention to the parentheses here.

$$(2^3)^2 = 2^3 \cdot 2^3 = 2^6,$$

$$\text{but } 2^{3^2} = 2^{3 \cdot 3} = 2^9$$

# How do you square parentheses?

If you square a sum, and expand it, you get two identical terms that you can combine.

$$(a + b)^2 = (a + b)(a + b) =$$
$$= a^2 \underbrace{+ab + ba} + b^2 =$$
$$= a^2 + 2ab + b^2$$

If you square a difference, it works in exactly the same way. Note that $(-b)(-b) = (-b)^2 = b^2$.

$$(a - b)^2 = (a - b)(a - b) =$$
$$= a(a - b) - b(a - b) =$$
$$= a^2 \underbrace{-ab - ba} + (-b)^2 =$$
$$= a^2 - 2ab + b^2$$

If you calculate $(a + b)(a - b)$, the $-ab$ cancels with $+ba$.

$$(a + b)(a - b) = a(a - b) + b(a - b) =$$
$$= a^2 - ab + ba + b(-b) =$$
$$= a^2 - b^2$$

You can also simply remember these as the three binomial formulas.

$$(a + b)^2 = a^2 + 2ab + b^2$$
$$(a - b)^2 = a^2 - 2ab + b^2$$
$$(a + b)(a - b) = a^2 - b^2$$

It's important that you do not forget the middle term in the first and second binomial formulas; the sign is the same as in the bracket. In both cases, there is a plus in front of $b^2$.

$$(a + b)^2 = a^2 + 2ab + b^2$$
$$(a - b)^2 = a^2 - 2ab + b^2$$

here always
minus plus

In the third binomial formula, there is no mixed term, but there is a minus in front of $b^2$.

$$(a + b)(a - b) = a^2 - b^2$$

minus

The binomial formulas are sometimes also helpful when calculating with numbers.

$$31 \cdot 29 = (30 + 1) \cdot (30 - 1) =$$
$$= 30^2 - 1 = 900 - 1 = 899$$

$$1.01^2 =$$
$$= 1^2 + 2 \cdot 1 \cdot 0.01 + 0.01^2 = 1.0201$$

# What are roots for and how do you calculate with them?

If you have a square and want to calculate the base, you need the square root, which is also simply referred to as root.

$$3^2 = 9, 3 = \sqrt{9}$$
$$x^2 = a, x = \sqrt{a}$$

The square root corresponds to raising to the power of $\frac{1}{2}$. Raising to the power of $\frac{1}{2}$ times raising to the power of $\frac{1}{2}$ gives raising to the power of $\frac{1}{2} + \frac{1}{2}$, that is, to the power of 1.

$$\sqrt{a} = a^{\frac{1}{2}},$$
because $\sqrt{a} \cdot \sqrt{a} =$
$$= a^{\frac{1}{2}} \cdot a^{\frac{1}{2}} = a^{\frac{1}{2} + \frac{1}{2}} = a^1 = a.$$
Therefore, the square root multiplied with itself gives the original number.

If you have a third power and want to calculate the base, you need the cube root.

$$3^3 = 27, \quad 3 = \sqrt[3]{27}$$
$$x^3 = a, \quad x = \sqrt[3]{a}$$

If you have an $n^{\text{th}}$ power, you need the $n^{\text{th}}$ root.

$$x^n = a, x = \sqrt[n]{a}$$

With roots you can calculate using the laws of exponents: you just have to rewrite the root as a power.

$$\sqrt{a} = a^{\frac{1}{2}}, \sqrt[n]{a} = a^{\frac{1}{n}}$$

If you raise a root to a power or take the root of a power, the exponent becomes a fraction.

$$(\sqrt[m]{a})^n = (a^{\frac{1}{m}})^n = a^{\frac{n}{m}}$$
$$\sqrt[m]{a^n} = (a^n)^{\frac{1}{m}} = a^{\frac{n}{m}}$$

Sometimes you can simplify, and you are left with only a root or only a power.

$$(\sqrt[4]{a})^2 = a^{\frac{2}{4}} = a^{\frac{1}{2}} = \sqrt{a}$$
$$(\sqrt{a})^4 = a^{\frac{4}{2}} = a^2$$
$$\sqrt[4]{a^2} = a^{\frac{2}{4}} = a^{\frac{1}{2}} = \sqrt{a}$$
$$\sqrt{a^4} = a^{\frac{4}{2}} = a^2$$

N.B. If you have a sum under the root, you must not take the root of each term separately.

$$\sqrt{3^2 + 4^2} \neq 3 + 4$$
$$\sqrt{25} = 5 \neq 7$$

# What is the absolute value of a number?

The absolute value of a number is always positive; it gives the distance of the number from 0.

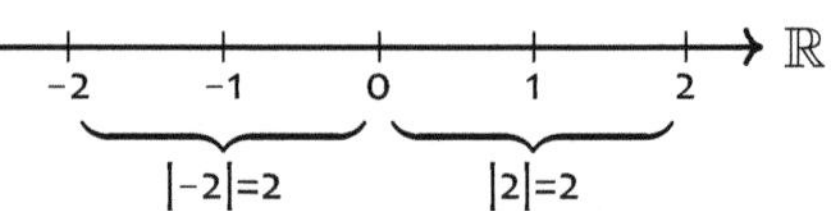

The absolute value is the number without its sign, i.e. if the number is positive, the absolute value is simply the number; if the number is negative, it's the negative of the number.

$$|a| = \begin{cases} a & \text{for} & a \geq 0 \\ -a & \text{for} & a < 0 \end{cases}$$

$$|2| = 2$$

$$|-2| = -(-2) = 2$$

Since the square root is defined as positive, the absolute value of a number is the square root of the square of that number.

$$|a| = \sqrt{a^2}$$

The set of all numbers whose absolute value is less than a positive number is an interval symmetric about 0.

$$(-2, 2) = \{x \mid -2 < x < 2\} = \{x \mid |x| < 2\}$$

The same applies also for all numbers that have a distance less than $\varepsilon$ from a number $x_0$.

$$(x_0 - \varepsilon, x_0 + \varepsilon) = \{x \mid x_0 - \varepsilon < x < x_0 + \varepsilon\}$$
$$= \{x \mid |x - x_0| < \varepsilon\}$$

This is called the $\varepsilon$-neighborhood of $x_0$.

$$U_\varepsilon(x_0) = \{x \mid |x - x_0| < \varepsilon\}$$

The absolute value of a sum is at most as large as the sum of the absolute values. Equality holds if $a$ and $b$ have the same sign.

$$|a + b| \leq |a| + |b|$$

The absolute value of a sum is at least as large as the absolute value of the difference of the absolute values.

$$\bigl||a| - |b|\bigr| \leq |a + b|$$

# Why do you need logarithms?

If you have, for example, a power of ten and want to calculate the exponent, you need the common logarithm, denoted by $\log_{10}$ or simply log.

$$\log(10^1) = \log(10) = 1$$
$$\log(10^2) = \log(100) = 2$$
$$\log(10^3) = \log(1000) = 3$$

The common logarithm in a sense counts the number of zeros after the 1.

$$\log(10) = 1$$
$$\log(100) = 2$$
$$\log(1000) = 3$$

This works similarly for negative powers of ten.

$$\log(10^{-1}) = \log(0.1) = -1$$
$$\log(10^{-2}) = \log(0.01) = -2$$
$$\log(10^{-3}) = \log(0.001) = -3$$

The logarithm of 1 is 0.

$$\log(1) = 0, \text{ since } 10^0 = 1$$

It's also defined for all other positive numbers. To find the common logarithm of 2, we look for a number $x = \log(2)$ such that $10^x = 10^{\log(2)} = 2$.

$$10^{\log(2)} = 2$$
$$10^0 = 1 < 2$$
$$10^1 = 10 > 2$$
$$\Rightarrow 0 < \log(2) < 1$$

We can try out different numbers and thus bracket $x$.

$$10^{0.25} = \sqrt[4]{10} \approx 1.78 < 2$$
$$10^{0.33\cdots} = \sqrt[3]{10} \approx 2.15 > 2$$
$$\Rightarrow 0.25 < \log(2) < 0.33\ldots$$

You can determine $\log(2)$ to any desired accuracy by constructing a sequence of nested intervals containing it.

$$10^{0.30102} \approx 1.999954$$
$$10^{0.30103} \approx 2.00000002$$
$$\Rightarrow \log(2) = 0.30102\ldots$$

# What is a logarithm to a base other than 10?

If you have a power of two and want to determine the exponent, you need the base-2 logarithm $\log_2$.

$$\log_2(2) = \log_2(2^1) = 1$$
$$\log_2(4) = \log_2(2^2) = 2$$
$$\log_2(8) = \log_2(2^3) = 3$$

You can easily convert between different logarithms; for this, you need to know $\log(a^b) = b\log(a)$.

$$2^x = c \quad | \text{ take logarithms}$$
$$x\log(2) = \log(c) \quad | \div \log(2)$$
$$x = \frac{\log(c)}{\log(2)}$$

On the other hand: $x = \log_2(c)$.

Thus, $\log_2(c) = \frac{\log(c)}{\log(2)}$.

Therefore, you actually only need one logarithm. In calculus, the logarithm to the base $e = 2.71828\ldots$ is the most practical, since its derivative is $\frac{1}{x}$.

$$\log_e(x) = \frac{\log(x)}{\log(e)}$$

It's called the natural logarithm $\ln(x)$.

$$\ln(x) = \log_e(x)$$

# What does it mean to write logarithm as log and not as an operation symbol?

We see here for the first time that a function is not expressed by an operation symbol, like the root, but by an abbreviated term, "log" for "logarithm".

log(2): logarithm of 2

In programming languages the root is also written as an abbreviated term, "sqrt" for "square root".

$\sqrt{2}$ = sqrt(2): square root of 2

By the way, the root symbol originated from the letter r for radix (Latin for root); the right arc of the r became the roof of the root symbol.

Later, we will encounter many more functions that are written out.

$\sin(45°)$ for sine of $45°$
$\exp(x)$ for $e^x$

To make it clear that the function acts on the argument, and is not a variable, parentheses are used. This also demands that you first compute the function before continuing with further calculations.

log(2)
↑
argument

Sometimes the parentheses are left out. However, this is not recommended.

Some people write
log 2 instead of log(2)
sin 45° instead of sin(45°).

# How do you calculate with logarithms?

From the laws of exponents, the laws of logarithms follow directly. When calculating the logarithm of the product of two numbers, you must add their logarithms.

$$a \cdot b = 10^{\log(a)} \cdot 10^{\log(b)}$$
$$a \cdot b = 10^{\log(a)+\log(b)}$$
$$\Rightarrow \log(a \cdot b) = \log(a) + \log(b)$$

When calculating the logarithm of the power of a number, you must multiply the logarithm of the number by the exponent.

$$a^b = (10^{\log(a)})^b$$
$$a^b = 10^{b \cdot \log(a)}$$
$$\Rightarrow \log(a^b) = b \cdot \log(a)$$

A "plus" in the exponent becomes "times", a "minus" becomes "divided by". The operation "times" in the exponent becomes "to the power of".

$$2^{3+5} = 2^3 \cdot 2^5$$
$$2^{3-5} = 2^3/2^5$$
$$2^{3 \cdot 5} = (2^3)^5$$

When taking logarithms, it's the other way around: "times" becomes "plus", "divided by" becomes "minus", "to the power of" becomes "times".

$$\log(6) = \log(2 \cdot 3) = \log(2) + \log(3)$$
$$\log(\tfrac{2}{3}) = \log(2/3) = \log(2) - \log(3)$$
$$\log(8) = \log(2^3) = 3 \cdot \log(2)$$

To solve an equation with the unknown in the exponent, you take the logarithm of both sides; then the unknown is at the bottom and you can solve for it using the basic arithmetic operations.

$$3^x = 9$$
$$x \log(3) = \log(9)$$
$$x = \frac{\log(9)}{\log(3)} = 2$$

You can easily convert between different logarithms.

$$b^x = c \quad (*)$$
gives, by the definition of $\log_b$:
$$x = \log_b(c).$$

Taking logarithms of $(*)$:
$$x \log(b) = \log(c)$$

$$\Rightarrow \log_b(c) = \frac{\log(c)}{\log(b)}$$

# Why are there two inverse operations for exponentiation?

Have you ever wondered why multiplication has only one inverse (division), but exponentiation has two, namely roots and logarithms?

This is because in multiplication, both factors are treated equally, whereas in exponentiation, the base and the exponent are fundamentally different.

Therefore, you need one inverse operation (the root) to solve for the base, and another inverse operation (the logarithm) to solve for the exponent.

Each of the two inverses uses the other number as a parameter.

$3 \cdot 4 = 4 + 4 + 4 = 12$
$4 \cdot 3 = 3 + 3 + 3 + 3 = 12 = 12 = 3 \cdot 4$

$3^4 = 3 \cdot 3 \cdot 3 \cdot 3 = 81$
$4^3 = 4 \cdot 4 \cdot 4 = 64 \neq 81 = 3^4$

$3^4 = 81$

$\sqrt[4]{81} = 3$ (solving for the base)
$\log_3(81) = 4$
(solving for the exponent)

fourth root: $\sqrt[4]{\ldots}$
logarithm to base three: $\log_3(\ldots)$

# Overview: Exponents, roots, logarithms

The three binomial formulas help you with squaring and expanding parentheses.

$$(a + b)^2 = a^2 + 2ab + b^2$$
$$(a - b)^2 = a^2 - 2ab + b^2$$
$$(a + b)(a - b) = a^2 - b^2$$

Exponents with the same base are multiplied by adding the exponents.

$$a^n \cdot a^m = a^{n+m}$$

Exponents with the same base are divided by subtracting the exponents.

$$\frac{a^n}{a^m} = a^{n-m}$$

Multiplying by 1 changes nothing, just as adding 0 does not. Therefore, $a^0 = 1$.

$$a^0 = 1$$

Raising to the power of −1 is the same as dividing 1 by the base. Raising $a$ to the power of −$n$ is the same as one over $a$ raised to the power of $n$.

$$a^{-1} = \frac{1}{a}, \quad a^{-n} = \frac{1}{a^n}$$

To raise a power to another power, you multiply the exponents.

$$(a^b)^c = a^{bc}$$

The $n^{th}$ root is the same as raising to the $(1/n)$-th power.

$$\sqrt[n]{a} = a^{1/n}$$

The logarithm of a product is the sum of the logarithms.

$$\log(a \cdot b) = \log(a) + \log(b)$$

The logarithm of a fraction is the difference of the logarithms.

$$\log\left(\frac{a}{b}\right) = \log(a) - \log(b)$$

The logarithm of a power is the exponent times the logarithm.

$$\log(a^n) = n \log(a)$$

# Some Propositional Logic and Set Theory

In mathematics, it's crucial to know whether an equation or, more generally, a statement is true or false.

You can deduce the truth value of a composed statement from the truth values of its parts; for more complicated statements, this is called proving.

In the following, you will learn how to calculate with "true" and "false" and what it means when rearrangements do not change the truth value.

Often, equations depend on variables. They are true for certain values of the variables, and false for others.

The values of the variables for which an equation is true form the solution set of the equation.

You will learn how to calculate with sets and how this relates to the operations on truth values.

Finally, I will show you different proof methods by working through an example.



A. Gründers, *Math Made Clear: From the Basics to Calculus*, https://doi.org/10.1007/978-3-662-73221-2_3

# How do you calculate with "true" and "false"?

Mathematical statements can be true or false. Instead of "true", you can also say "correct".

$3 + 2 = 5$ is true.
$4 > 6$ is false.

You can also write this as an equation.

$A = $ "$3 + 2 = 5$" $= T$
$B = $ "$4 > 6$" $= F$

If you negate a statement $A$, you write $\overline{A}$ ("not $A$"). $\overline{A}$ is true if $A$ is false, and false if $A$ is true.

| $A$ | $\overline{A}$ |
|---|---|
| T | F |
| F | T |

$\overline{A} = $ "$3 + 2 \neq 5$" $= F$

You can combine statements and introduce symbols as abbreviations for this purpose.

$C = A \wedge B$
$D = A \vee B$

$A$ and $B$ is written as $A \wedge B$. This statement is only true if both $A$ and $B$ are true. Otherwise, it's false.

| $A$ | $B$ | $A \wedge B$ |
|---|---|---|
| T | T | T |
| T | F | F |
| F | T | F |
| F | F | F |

"$3 + 2 = 5$" $\wedge$ "$4 > 6$" $= F$

$A$ or $B$ is written as $A \vee B$. This statement is true if either $A$ or $B$ is true, or both are true. Otherwise, it's false.

| $A$ | $B$ | $A \vee B$ |
|---|---|---|
| T | T | T |
| T | F | T |
| F | T | T |
| F | F | F |

"$3 + 2 = 5$" $\vee$ "$4 > 6$" $= T$

# What does "$A$ implies $B$" mean?

If a statement depends on a variable, it's called a propositional function or a statement form or just a statement.

$A(x) =$ "$2x + 1 = 3$"

In general, a propositional function $A(x)$ is true for a certain set of $x$ and false for all other $x$.

$A(x)$ is true for $x = 1$,
it's false for all other $x$.

In mathematics we often want to deduce other statements from given statements.

We want to deduce a statement of the form "$x = \ldots$" from $A(x)$ above.

For the statement "$A$ implies $B$" we write $A \Rightarrow B$. It's false if $A$ is true and $B$ is false. Otherwise, it's true. In particular, if $A$ is false, then the implication is true regardless of the truth value of B.

| A | B | $A \Rightarrow B$ |
|---|---|---|
| T | T | T |
| T | F | F |
| F | T | T |
| F | F | T |

It may seem unusual that "$F \Rightarrow T$" is true, that is, if something true follows from something false, the whole statement is considered true.

The crucial point is that it's about the overall statement, i.e. the conclusion. The conclusion itself can only be false if the first statement is true (and the second false). Then the conclusion produces something false. If the first statement is already false, the second statement is no longer relevant. The conclusion itself can't therefore be false.

Thus, from a false statement you can derive everything, be it true or false. From a true statement, however, you can only derive true statements.

Something false can only arise if $A$ is true (and B is false).

$A \Rightarrow B$ is only false,
if $A$ is true and B is false.

# What do "necessary" and "sufficient" mean?

As an example, let us consider the multiplication of natural numbers and how it behaves with even and odd numbers.

| $n$ | $m$ | $n \cdot m$ | example |
|---|---|---|---|
| e | e | e | $4 \cdot 4 = 16$ |
| e | o | e | $4 \cdot 3 = 12$ |
| o | e | e | $3 \cdot 4 = 12$ |
| o | o | o | $3 \cdot 3 = 9$ |

We define
A = "first factor is even" and
B = "product is even".

| A | B |
|---|---|
| T | T |
| F | T/F |

Then it follows from statement A that statement B holds.

The product of an even number with another natural number is always even. It holds:
$n$ is even $\Rightarrow n \cdot m$ is even.

But from statement B, statement A does not follow.

The product can also be even if the first factor is not even. It does not hold:
$n \cdot m$ is even $\Rightarrow n$ is even.
$3 \cdot 4$ is even, $3$ is not even.

We call A "sufficient" for B. One could also say "enough", since if A holds, that is enough for B to hold. However, the term "sufficient" has become more established.

It's sufficient that the first factor is even for the product to be even.

A is not necessary for B to hold.

The product can also be even without the first factor being even.

B however is necessary for A.

If the product is not even, then the first factor cannot be even either, so the truth of B is necessary for the truth of A.

# What does "if and only if" mean?

It's very important to distinguish between $A \Rightarrow B$ and $B \Rightarrow A$.

If both $A \Rightarrow B$ and $B \Rightarrow A$ hold, we write $A \Leftrightarrow B$.

This means, that A and B have the same truth values. The connective $A \Leftrightarrow B$ is therefore called equivalence. We also say: A holds if and only if (abbreviated "iff") B holds. $A \Leftrightarrow B$ is true if and only if A and B have the same truth values.

"First factor even $\Rightarrow$ product even" is true.
"Product even $\Rightarrow$ first factor even" is false.

| A | B | $A \Leftrightarrow B$ |
|---|---|---|
| T | T | T |
| T | F | F |
| F | T | F |
| F | F | T |

When rearranging equations we want the equation to be true for certain values of the variables after the rearrangement if and only if it was true for those values before the rearrangement.

The solution set should stay unchanged, i.e. no spurious solutions should appear and no real solution should disappear.
Such a rearrangement is called an equivalence rearrangement.

Dividing by 2 is an equivalence rearrangement.

$2x = 6 \Leftrightarrow x = 3$
Left side is true exactly for $x = 3$.
Right side is true exactly for $x = 3$.

Squaring is not an equivalence rearrangement.

$x = 2 \Rightarrow x^2 = 4$
Left side is true exactly for $x = 2$.
Right side is true for $x = 2$ and $x = -2$.
A spurious solution has appeared.
$x = 2 \nLeftarrow x^2 = 4$
$x = \pm 2 \Leftrightarrow x^2 = 4$

# What is a set?

A set is the collection of well-distinguished objects into a whole. You mentally put all the objects into a bag and consider the bag with the elements.

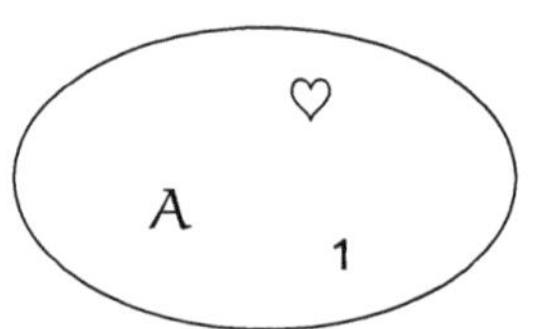


The objects are called elements, and the bag with the elements is called a set. The set is written with curly braces.

$\{A, 1, \heartsuit\}$    $\{A, 1, \heartsuit\}$

These are the elements.    This is the set.

We say, an element is contained in a set and write $\in$.

$1 \in \{A, 1, \heartsuit\}$

You can also denote a set by a letter.

$M = \{A, 1, \heartsuit\}$
$1 \in M$

For a set the order does not matter. Each element appears only once.

$\{1, 2, 3\} = \{3, 1, 2\}$

Two sets are equal if and only if they contain the same elements.

The number of elements in a set is called the cardinality and is denoted by two vertical bars.

$|\{A, 1, \heartsuit\}| = |\{1, 2, 3\}| = 3$

If a set contains no element, it's called the empty set and is sometimes denoted by $\emptyset$.

$\{\} = \emptyset$

The set with one element is different from the element itself.

$\{17\} \neq 17$

# How do you calculate with sets?

The intersection of two sets is the set that contains exactly the elements that are contained in the first set and in the second.

$A = \{2, 4, 6\}$
$B = \{5, 6\}$
$A \cap B = \{x \mid x \in A \wedge x \in B\}$
$A \cap B = \{6\}$

The union is the set that contains exactly the elements that are contained in the first set or in the second or in both.

$A \cup B = \{x \mid x \in A \vee x \in B\}$
$A \cup B = \{2, 4, 5, 6\}$

The difference set contains the elements of the first set that are not contained in the second.

$A \setminus B = \{x \mid x \in A \wedge x \notin B\}$
$A \setminus B = \{2, 4\}$

You can see that the set operations are closely connected to the logical operations. This is why their symbols are similar.

$\wedge$ and $\quad \cap$ intersection with
$\vee$ or $\quad\quad \cup$ union with

Set operations come with their own rules for calculation, but we will skip the deep dive. You may notice that there are two distributive laws.

$A \cap B = B \cap A, \quad\quad A \cup B = B \cup A$
$A \cap (B \cap C) = (A \cap B) \cap C = A \cap B \cap C$
$A \cup (B \cup C) = (A \cup B) \cup C = A \cup B \cup C$
$A \cap (B \cup C) = (A \cap B) \cup (A \cap C)$
$A \cup (B \cap C) = (A \cup B) \cap (A \cup C)$

When you pair each element of one set with every element of another, the resulting set of ordered pairs is called the Cartesian product. The cardinality of the Cartesian product equals the product of the cardinalities of the original sets.

| | 5 | 6 |
|---|---|---|
| 2 | $(2, 5)$ | $(2, 6)$ |
| 4 | $(4, 5)$ | $(4, 6)$ |
| 6 | $(6, 5)$ | $(6, 6)$ |

$\{2, 4, 6\} \times \{5, 6\} =$
$\{(2, 5), (2, 6), (4, 5), (4, 6), (6, 5), (6, 6)\}$

You write ordered pairs with round brackets. In round brackets, order matters. In curly bracket which are used for sets, order doesn't matter.

$(2, 5) \neq (5, 2)$ pairs
$\{2, 5\} = \{5, 2\}$ two-element sets

# What are intervals?

An interval is a set of real numbers that forms a segment on the number line.

$$(2, 3) = \{x \mid 2 < x < 3\}$$

Depending on whether the left or right boundary of the interval is included, we talk about closed, open, or half-open intervals. Where the boundary is included, we use square brackets, otherwise round brackets.

open:
$$(2, 3) = \{x \mid 2 < x < 3\}$$
half-open:
$$(2, 3] = \{x \mid 2 < x \leq 3\}$$
$$[2, 3) = \{x \mid 2 \leq x < 3\}$$
closed:
$$[2, 3] = \{x \mid 2 \leq x \leq 3\}$$

There are also unbounded intervals. These are shown as open on the side that goes to infinity. This is because $\infty$ or $-\infty$ is not a real number and isn't part of the interval.

$$(-\infty, 0] = \{x \mid x \leq 0\}$$
$$(-\infty, 0) = \{x \mid x < 0\}$$
$$[0, \infty) = \{x \mid 0 \leq x\}$$
$$(0, \infty) = \{x \mid 0 < x\}$$

If you take the intersection of two intervals, the result is another interval.

$$(2, 4) \cap (3, 5) = (3, 4)$$
$$(2, 4) \cap [3, 5) = [3, 4)$$

The symbol $\infty$ means infinity; it's not a real number. We'll explore this more when we discuss limits.

$\infty$ is not a real number.

$\infty$ stands symbolically for "greater than any arbitrarily large positive real number".
$-\infty$ stands symbolically for "less than any arbitrarily large negative real number".

# What are the most important types of mathematical proofs?

We want to show in three different ways that statement $A$ implies statement $B$, i.e. that the implication $A \Rightarrow B$ holds.

1. Direct proof: We start with assumption $A$ and try to derive conclusion $B$.

2. Indirect proof: The implication $A \Rightarrow B$ is equivalent to the contrapositive $\overline{B} \Rightarrow \overline{A}$, as can be seen from the truth table. So we start from the negated conclusion $\overline{B}$ and try to derive the negated assumption $\overline{A}$.

3. Proof by contradiction: The implication $A \Rightarrow B$ is equivalent to the fact that the negated conclusion $\overline{B}$ and $A$ are in contradiction, i.e. that $\overline{B} \wedge A = F$.

In this simple example, all three proof types can be carried out without difficulty. For more complex statements, often only the 2nd or the 3rd way are feasible.

We show $A \Rightarrow B$ where
$A$: $n$ is odd,
$B$: $n^2$ is odd.

$A$: $n$ odd
$n$ can be written as $n = 2k + 1$.
$\Rightarrow n^2 = 4k^2 + 4k + 1 = 2(2k^2 + 2k) + 1$
So $n^2$ is odd and $B$ is shown.

$\overline{B}$: $n^2$ even
$n^2$ can be written as $n^2 = 2m$.
$\Rightarrow n$ must contain the factor 2.
So $n$ is even and $\overline{A}$ is shown.

$n^2 = 2k \quad (*)$
$n = 2l + 1$
$n^2 = 4l^2 + 4l + 1 \quad (**)$
$(**) - (*) : 0 = 2(l^2 + 2l - k) + 1$
On the right is an odd number, on the left the even number 0.
So we get the desired contradiction.

The challenge with the 3rd method is that you must derive a contradiction, i.e. you are dealing with false statements, but these must result from $\overline{B} \wedge A$ and not from a mistake.

# Solving Equations for Unknowns

Solving equations for unknown quantities is a fundamental method, especially for applications.

You are given an equation or you have set one up based on a word problem, and you are looking for the value of an unknown quantity, often denoted by $x$.

The set of all $x$ values for which the equation is true when substituted into the equation is the solution set, i.e. the set of all solutions to the problem.

Often there is exactly one solution, but this does not have to be the case.

Equations with unknowns are one way how mathematics expresses questions. The answer to the question is the value of the unknown.

In the previous chapter, we saw what the truth value of a statement is, and that equivalent statements have the same truth value.

Equivalence rearrangements are therefore useful for solving equations, as they do not change truth values, and thus don't change the solution set either.

© The Author(s), under exclusive license to Springer-Verlag GmbH, DE, part of Springer Nature 2026
A. Gründers, *Math Made Clear: From the Basics to Calculus*,
https://doi.org/10.1007/978-3-662-73221-2_4

# How do you rearrange equations using equivalence rearrangements?

When you rearrange equations to solve for an unknown, you want to achieve two things.

First, the equation should be true for certain values of the variable after the rearrangement if and only if the equation before was true for these values.

$2x + 1 = 7$ is true for exactly the same values of $x$ as the equation $x = 3$.

In other words, no solutions should be added or lost.

If this is the case, the rearrangement is called an equivalence rearrangement and we write $\Leftrightarrow$. The rearranged equation is then equivalent to the original one.

$2x + 1 = 7 \Leftrightarrow x = 3$
The two equations are equivalent.

Second, after the rearrangement, the unknown should appear only on one side of the equation, and alone. Then you can read off the value on the other side as the solution.

$x = 3$
On the left side $x$ stands alone, on the right side you can see the value of $x$, the solution.

Usually you cannot achieve this in one step, but must perform several equivalence rearrangements.

$2x + 1 = 7 \quad | - 1$
$\Leftrightarrow 2x = 6 \quad | \div 2$
$\Leftrightarrow x = 3$
Subtracting and dividing on both sides are equivalence rearrangements.

# What are typical equivalence rearrangements?

You can add or subtract a number on both sides.

$$x + 7 = 5 \quad | - 7$$
$$x = 5 - 7 = -2$$

You can multiply or divide both sides by a number.

$$7x = 5 \quad | \div 7$$
$$x = \frac{5}{7}$$

The number must be nonzero, since multiplication by 0 is not an equivalence rearrangement, and division by 0 is not defined.

Multiplying by 0 is not an equivalence rearrangement: multiplying the false statement $3 = 5$ by 0 leads to the true statement $0 = 0$.

If you multiply by an expression that can become zero depending on $x$, you must exclude the corresponding values of $x$.

$$\frac{x - 1}{x^2 - 1} = 1 \quad | \cdot (x^2 - 1) \neq 0, \text{ so } x \neq \pm 1$$
$$x - 1 = x^2 - 1 \quad | + 1$$
$$x = x^2$$
$$x(1 - x) = 0$$
$$\Rightarrow x_1 = 0, x_2 = 1$$
$$x = 0, \text{ since } x_2 = 1 \text{ is excluded.}$$

You can take the logarithm of both sides. This helps when the unknown is in the exponent.

$$2^x = 5 \quad | \log$$
$$\log(2^x) = \log(5)$$
$$x \log(2) = \log(5) \quad | \div \log(2)$$
$$x = \frac{\log(5)}{\log(2)}$$

When you take the square root, remember that you get two solutions (this applies to all even roots).

$$x^2 = 9$$
$$x = \pm\sqrt{9} = \pm 3$$
$$x_1 = -3, x_2 = +3$$

If a product is equal to 0, it follows that at least one of the factors must be 0.

$$(x - 1)(e^x - 2) = 0$$
$$\Rightarrow x - 1 = 0 \text{ or } e^x - 2 = 0$$
$$\Rightarrow x_1 = 1, x_2 = \ln(2)$$

# Where do you have to be careful when rearranging equations?

You must always be careful when the rearrangements are not equivalence rearrangements.

Multiplying by a number is only an equivalence rearrangement if you multiply by a number not equal to zero.

Squaring an equation implies the squared equation, but the squared equation does not imply the original equation.

It may happen that you have to perform rearrangements that are not equivalence rearrangements, e.g. when solving radical equations. You must then always check your solutions by substituting them back into the original equation.

Dividing by zero is not only not an equivalence rearrangement, but not defined. If you do it anyway, you can "prove" false statements such as 1 = 0.

Squaring and multiplying by zero are not equivalence rearrangements.

When multiplying by 0, a false equation like $3 = 5$ becomes the true statement $0 = 0$. Therefore, it's not an equivalence rearrangement.

$$x = 1 \Rightarrow x^2 = 1$$
$$x = 1 \nLeftarrow x^2 = 1,$$
The left side is true only for $x = 1$.
The right side is true for $x = 1$ and $x = -1$.
By squaring, a spurious solution has been introduced.

$$\sqrt{x + 1} + \sqrt{x - 1} = 1 \quad | \text{ squaring}$$
$$x + 1 + 2\sqrt{x + 1}\sqrt{x - 1} + x - 1 = 1$$
$$2x + 2\sqrt{x + 1}\sqrt{x - 1} = 1 \quad | - 2x$$
$$2\sqrt{x + 1}\sqrt{x - 1} = 1 - 2x$$
$$4(x^2 - 1) = 1 - 4x + 4x^2 \quad | - 1$$
$$-5 = -4x \quad | \div (-5)$$
$$x = \frac{5}{4}$$
Check: $\sqrt{\frac{9}{4}} + \sqrt{\frac{1}{4}} = \frac{3}{2} + \frac{1}{2} = 2 \neq 1$
The equation has no solution.

Set $x = 1$ (this is allowed).
$$x = 1 \quad | \cdot x \quad \text{(correct)}$$
$$x^2 = x \quad | - x \quad \text{(correct)}$$
$$x^2 - x = 0 \quad | \text{ 3rd binomial formula}$$
$$(x+1)(x-1) = 0 | \div (x-1) \text{ (not defined!)}$$
$$x + 1 = 0 \quad | \text{ substitute } x = 1$$
$$2 = 0 \quad | \div 2 \text{ (allowed)}$$
$$1 = 0$$

# How do you recognize the type of an equation?

Check first how many unknowns your equation contains. We will deal with the case of several unknowns later.

$x + \sin(x) = 0$:    1 unknown: $x$
$x^2 + y^2 = 3$:      2 unknowns: $x, y$

Then check whether you can simplify the equation in an obvious way.

$$x^2 + x = x + 1 \Rightarrow x^2 = 1$$

Bring the fractions to a common denominator.

$$\frac{1}{x + 1} + \frac{1}{x - 1} = 1 \Rightarrow \frac{2x}{x^2 - 1} = 1$$

Check whether the unknown appears only linearly. Then you have a linear equation.

$$\frac{17}{35}x + \frac{101}{1001} = 717$$
see linear equation

If the unknown appears squared and possibly linear, you have a quadratic equation.

$$x^2 + x + 1 = 0$$
see quadratic equation

If the unknown appears as a higher power, you have a polynomial equation of higher degree, and you need to find the zeros of a polynomial.

$$x^3 + x^2 + x = 0$$
see higher degree equations

If the equation contains roots, you must cleverly rearrange and square both sides to eliminate the roots. You must always check your solutions, as squaring can introduce spurious solutions.

$$\sqrt{x + 1} + \sqrt{x - 1} = 1$$
see root equation under equivalent rearrangement

If the unknown appears in the exponent, you have an exponential equation.

$$4^x + 2^x = 5$$
see exponential equation

# How do you solve a linear equation?

You want to solve a linear equation in one variable.

$$3x + 2 = 7$$

To solve the equation, you need to isolate $x$. To do this, first move the term without $x$ to the other side, i.e. subtract 2 from both sides.

$$3x = 7 - 2 = 5$$

To have $x$ standing alone on one side, you still need to get rid of the 3 on the left. You do this by dividing both sides by 3.

$$3x = 5 \quad | \div 3$$
$$x = \frac{5}{3}$$

In exactly the same way, you can solve any linear equation (with $a \neq 0$, otherwise it's not a linear equation) in one variable.

$$ax + b = c$$
$$x = \frac{c - b}{a}$$

But you don't need to memorize the formulas, as they result automatically if you proceed as above.

# How do you solve a quadratic equation?

Some quadratic equations can be solved by simply taking the square root of both sides.

$$x^2 = 4$$
$$x_{1,2} = \pm\sqrt{4} = \pm 2$$

To clarify that there are two solutions, we write $x_{1,2}$. The $x_1$ is the solution with +, $x_2$ is the solution with −.

$$x_{1,2} = \pm 2$$
stands for
$$x_1 = +2$$
$$x_2 = -2$$

The notable thing about this quadratic equation was that $x$ only appeared as $x^2$.

If you have a more generic quadratic equation, the solution is quite similar: you also need to take a square root, and in general you will also get two solutions.

$$x^2 - 2x = 5$$

But you need a step before.

The trick to get rid of the linear term with $x$ is called completing the square.

Recall the binomial formula (or multiply it out).

$$(a - b)^2 = a^2 - 2ab + b^2, \text{ so}$$
$$(x - 1)^2 = x^2 - 2x + 1$$

Add 1 to both sides so that you can apply the binomial formula on the left hand side.

$$x^2 - 2x = 5$$
$$\Rightarrow x^2 - 2x + 1 = 5 + 1 = 6$$
$$\Rightarrow (x - 1)^2 = 6$$

Then simply take the square root on both sides, just as above.

$$\Rightarrow x_{1,2} - 1 = \pm\sqrt{6}$$

Now you just need to get rid of the −1 on the left side. To do this, add 1 to both sides. The number is usually written in front of the root, since it's the simpler term.

$$\Rightarrow x_{1,2} = 1 \pm \sqrt{6}$$

# How does completing the square work in general?

If you carry out the procedure from the previous page for the quadratic equation $x^2 + px + q = 0$, you obtain the quadratic formula.

$$x^2 + px + q = 0$$

$$\left(x + \tfrac{p}{2}\right)^2 - \left(\tfrac{p}{2}\right)^2 + q = 0$$

$$\left(x + \tfrac{p}{2}\right)^2 = \left(\tfrac{p}{2}\right)^2 - q$$

$$x_{1,2} + \tfrac{p}{2} = \pm\sqrt{\left(\tfrac{p}{2}\right)^2 - q}$$

$$x_{1,2} = -\frac{p}{2} \pm \sqrt{\left(\frac{p}{2}\right)^2 - q}$$

If you have an additional factor in front of $x^2$, you must first divide by this factor, and then you can apply the quadratic formula.

$$ax^2 + bx + c = 0 \quad | \div a$$

$$x^2 + \tfrac{b}{a}x + \tfrac{c}{a} = 0$$

The resulting formula is also called the quadratic formula.

$$x_{1,2} = \frac{-b \pm \sqrt{b^2 - 4ac}}{2a}$$

You do not need to memorize these formulas, since you can quickly derive them from the given equation each time.

What is crucial is completing the square: add the term that turns the quadratic and linear terms into a perfect square, and then subtract it again.

$$x^2 - 6x + 3 = 0$$

$$x^2 - 6x + 9 - 9 + 3 = 0$$

The term to be added is half the coefficient of $x$, squared.

Coefficient of $-6x$ is $-6$, half of that is $-3$, the square is 9, so add $+9 - 9$.

You write the completed square as a squared linear term.

$$x^2 \underbrace{- 6x + 9}_{(x-3)^2} \underbrace{-9 + 3}_{-6} = 0$$

$$(x_{1,2} - 3)^2 = 6$$

$$x_{1,2} - 3 = \pm\sqrt{6} \Rightarrow x_{1,2} = 3 \pm \sqrt{6}$$

# How do you solve higher-degree equations?

For the solutions of cubic and quartic equations, there are similar formulas as for the quadratic equation, but much more complicated. So, in practice, the zeros are usually calculated numerically.

For the general equation of degree five ("quintic equation"), it can be proven that there are no formulas with nested radicals for the solutions.

Approaches you can try in simple cases:

- Factoring out $x$ or a power of $x$, and what remains is a polynomial of degree one or two.

- Guessing a solution as follows: if your polynomial has integer coefficients, an integer solution is always a divisor of the constant term, i.e the term in the polynomial without $x$. If substituting it into the polynomial gives 0, you have found a solution.

- Setting an auxiliary variable (e.g. $z$) equal to $x^2$ gives you a simpler equation, e.g. a quadratic equation.

$$x^3 + px + q = 0$$
$$x = \sqrt[3]{-\frac{q}{2} + \sqrt{D}} + \sqrt[3]{-\frac{q}{2} - \sqrt{D}}$$
$$D = \left(\frac{p}{3}\right)^3 + \left(\frac{q}{2}\right)^2$$

The cube roots must be chosen so that their product is $-\frac{p}{3}$.

$$x^5 - x + 1 = 0$$
has been proven to have no solution with nested radicals.

$$x^{10} - 2x^9 = 0$$
$$\Rightarrow x^9(x - 2) = 0$$
$$\Rightarrow x_1 = 0, x_2 = 2$$
are the two solutions.

$$x^3 + x^2 - 5x + 3 = 0$$
If there are integer solutions, they are divisors of 3, i.e., 1 or –1 or 3 or –3. By trial:
$x_1 = 1$ and $x_2 = -3$ are solutions.

Once you have a solution, you can use polynomial division to reduce the degree of the equation by one.

$$x^4 - 5x^2 + 6 = 0$$
set $x^2 = z$, so $x^4 = z^2$
$$\Rightarrow z^2 - 5z + 6 = 0$$
$$\Rightarrow z_1 = 3, z_2 = 2 \text{ (see above)}$$
With $x = \pm\sqrt{z}$ it follows:
$$x_1 = \sqrt{3}, x_2 = -\sqrt{3}$$
$$x_3 = \sqrt{2}, x_4 = -\sqrt{2}$$

# How do you solve exponential equations?

If the unknown appears in the exponent, then you apply the logarithm corresponding to the base.

$$2^x = 5$$
$$x = \log_2(5)$$

You can also apply a logarithm to any other base, then the exponent also comes down and you get a linear equation.

$$2^x = 5$$
$$\ln(2^x) = \ln(5)$$
$$x\,\ln(2) = \ln(5)$$
$$x = \frac{\ln(5)}{\ln(2)}$$

If the unknown of an exponent appears twice and one base is the square of the other, you can abbreviate one exponential expression with $z$ and obtain a quadratic equation for $z$.

$$4^x + 2^x = 5$$
$$2^{2x} - 5 \cdot 2^x + 6 = 0$$
$$(2^x)^2 - 5 \cdot 2^x + 6 = 0 \;|\; \text{set } z = 2^x$$
$$z^2 - 5z + 6 = 0 \;|\; \text{solve the equation}$$
$$z_1 = 2,\, z_2 = 3$$
$$2^{x_1} = 2,\, 2^{x_2} = 3$$
$$x_1 = 1,\, x_2 = \log_2(3)$$

This works if all the bases are powers of one of the bases.

$$27^x + 2 \cdot 9^x + 3^x = 48 \;|\; \text{Set } z = 3^x.$$
$$z^3 + 2z^2 + z = 48$$

You can similarly solve the equation $\frac{e^x - e^{-x}}{2} = a$. The function on the left hand side is also called the hyperbolic sine, denoted "sinh". it's related to the sine function.

$$\frac{e^x - e^{-x}}{2} = a \;\Big|\quad z = e^x \Rightarrow e^{-x} = \frac{1}{z}$$
$$z - \frac{1}{z} = 2a \quad |\cdot z = e^x > 0$$
$$z^2 - 2az - 1 = 0 \quad |\, \text{solve the equation}$$
$$z_{1,2} = a \pm \sqrt{a^2 + 1} \;|\; \text{since } z = e^x > 0:$$
$$x = \ln(z_1) = \ln(a + \sqrt{a^2 + 1})$$

If $x$ and $a^x$ or $x$ and $\sin(x)$ or $x$ and $\ln(x)$ appear together in an equation, the solution can't be represented as a formula with elementary functions, in general.

$$x + 2^x = 0$$
$$xe^x = 1$$
not analytically solvable,
only with numerical methods

# How do you determine the solution set of an absolute value equation?

To determine the solution of an absolute value equation, you must consider both cases separately. The first case is that the argument of the absolute value is positive, the second, that it's negative.

$$|x| = \begin{cases} x & \text{for} & x \geq 0 \\ -x & \text{for} & x < 0 \end{cases}$$

You first determine the set of all $x$-values for the case that the argument is greater than or equal to 0 and calculate $x$. You then proceed in the same manner for the second case where the argument of the absolute value is less than 0.

$|x + 2| = 3$

$1^{\text{st}}$ case: $x + 2 \geq 0$, i.e. $x \geq -2$
$|x + 2| = x + 2 = 3 \Rightarrow x = 1$,
$x = 1$ satisfies $x \geq -2$, so $x_1 = 1$.

$2^{\text{nd}}$ case: $x + 2 < 0$, i.e. $x < -2$
$|x + 2| = -(x + 2) = 3 \Rightarrow x = -5$,
$x = -5$ satisfies $x < -2$, so $x_2 = -5$.

At the end you collect all solutions.

$x_1 = 1, x_2 = -5$

If you have several absolute values, you must consider the corresponding cases.

$|x + 2| + 2|x - 2| = 10$

1.) $x < -2$ : $-(x + 2) - 2(x - 2) = 10$
$\Rightarrow x = -\frac{8}{3}$, satisfies the case

2.) $-2 \leq x < 2$ : $(x + 2) - 2(x - 2) = 10$
$\Rightarrow x = -4$, doesn't satisfy the case

3.) $2 \leq x$ : $(x + 2) + 2(x - 2) = 10$
$\Rightarrow x = 4$, satisfies the case

At the end you collect all solutions.

$x_1 = -\frac{8}{3}, x_2 = 4$

# How do you determine the solution set of an inequality?

You can solve an inequality in a similar way to an equation.

$$2x + 5 > 3 \mid -5 \qquad 2x + 5 = 3 \mid -5$$
$$2x > -2 \mid \div 2 \qquad x = -2 \mid \div 2$$
$$x > -1 \qquad\qquad x = -1$$

However, if you multiply or divide by a negative number, you must reverse the inequality sign.

$$-x > 3 \quad \mid \cdot (-1)$$
$$x < -3$$

If you have a quadratic inequality, it's best to first solve the corresponding quadratic equation.

$$x^2 - 5x + 6 > 0$$
$$x^2 - 5x + 6 = 0 \Rightarrow x_1 = 2, x_2 = 3$$

The solution set then consists of one or two intervals, the bounds of which are the solutions of the equation. You determine which solution to use by plugging in a sample value. The inequality sign changes at each simple zero of the equation.

For $x < 2$ (e.g., $x = 0$)
$$x^2 - 5x + 6 > 0$$
For $2 < x < 3$ we have $x^2 - 5x + 6 < 0$.
For $3 < x$ we have $x^2 - 5x + 6 > 0$.
Solution set: $x < 2$ or $x > 3$,
so $x \in (-\infty, 2) \cup (3, \infty)$

Alternatively, you can factor the quadratic term, which is supposed to be greater than zero, using its roots and considering the inequality sign.

$$x^2 - 5x + 6 = (x - 2)(x - 3) > 0$$

$$a \cdot b > 0$$
$$\Rightarrow (a > 0 \wedge b > 0) \vee (a < 0 \wedge b < 0)$$

a) $x - 2 > 0 \wedge x - 3 > 0$
$$\Rightarrow x > 2 \wedge x > 3 \Rightarrow x > 3$$

b) $x - 2 < 0 \wedge x - 3 < 0$
$$\Rightarrow x < 2 \wedge x < 3 \Rightarrow x < 2$$

So the solution set is $x < 2 \vee x > 3$

# How do you solve a system of linear equations?

Two linear equations with two unknowns are called a system of linear equations.

$$(1) \quad 2x + 3y = 5$$
$$(2) \quad 3x + 5y = 10$$

One way to solve it is to multiply the first equation by one number and the second by another, so that one variable is eliminated when you subtract the two equations from each other.

$$5 \cdot (1) \quad 10x + 15y = 25$$
$$3 \cdot (2) \quad 9x + 15y = 30$$
$$\Rightarrow (10 - 9) \cdot x + 0 \cdot y = 25 - 30$$
$$\Rightarrow x = -5$$

Plugging into (1) gives $y = 5$.

This also works in general.

$$(1) \quad ax + by = e$$
$$(2) \quad cx + dy = f$$

You multiply the first equation by the coefficient of $y$ of the second equation, i.e. $d$, and the second equation by the coefficient of $x$ of the first equation, i.e. $b$.

$$d \cdot (1) \quad ad\,x + bd\,y = de$$
$$b \cdot (2) \quad bc\,x + bd\,y = bf$$

Then you subtract the equations from each other and solve the linear equation for $x$.

$$(ad - bc) \cdot x + 0 \cdot y = de - bf$$
$$x = \frac{de - bf}{ad - bc}$$

The 2 × 2 scheme of the coefficients is called the coefficient matrix. It's written with round parentheses.

$$\begin{pmatrix} a & b \\ c & d \end{pmatrix}$$

The quantity $ad - bc$ is called the determinant of the coefficient matrix and is written with "det" or with vertical bars.

$$\det \begin{pmatrix} a & b \\ c & d \end{pmatrix} = \begin{vmatrix} a & b \\ c & d \end{vmatrix} = ad - bc$$

You can write the solution as the quotient of two determinants.

$$x = \frac{\begin{vmatrix} e & b \\ f & d \end{vmatrix}}{\begin{vmatrix} a & b \\ c & d \end{vmatrix}}$$

# A Glimpse into Geometry

Geometry is an independent branch of mathematics, one of its roots, and an active field of research.

Here I will only explain a bit of geometry, just as much as one should know and as is needed in common applications.

The focus is on the calculation of lengths, areas, angles, and not on geometric constructions.

We will look at triangles, since these are the simplest planar geometric figures and all straight-edged planar shapes can be decomposed into triangles.



A. Gründers, *Math Made Clear: From the Basics to Calculus*, https://doi.org/10.1007/978-3-662-73221-2_5

# Why do the angles in a triangle sum up to $180°$?

If you consider the base side of a triangle and draw a parallel through the opposite vertex, you can see that the sum of the angles, $\alpha + \beta + \gamma$, is equal to a straight angle, i.e. $180°$.

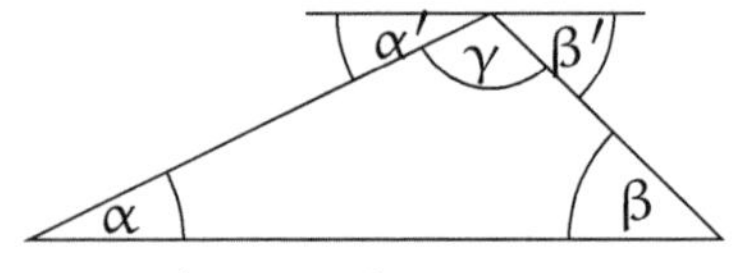

$$\alpha = \alpha', \beta = \beta'$$
$$\alpha + \beta + \gamma = \alpha' + \beta' + \gamma = 180°$$

$$\alpha + \beta + \gamma = 180°$$
$$\Rightarrow \gamma = 180° - (\alpha + \beta)$$

This means that if you are given two angles of a triangle, the third is determined: it's the difference between $180°$ and their sum.

You can also understand that the sum of the angles equals $180°$ as follows: if you place a pencil on one side of a triangle (bottom arrow), trace along the triangle, and at each vertex (right, top, left) turn it by the corresponding angle, then at the end the pencil points in the opposite direction (inner arrow), i.e. you have rotated it by $180°$.

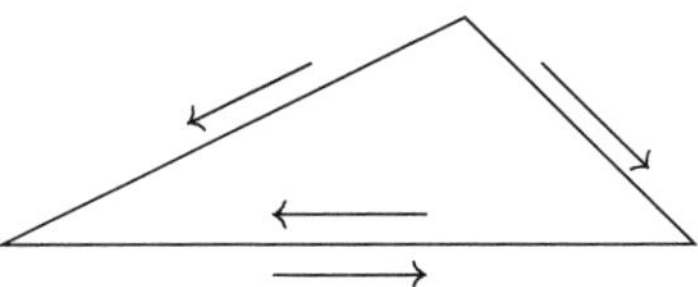

(1) Afterwards, the arrow points in the opposite direction.
(2) The total angle is less than $540°$, since each angle is less than $180°$.

(1) and (2) $\Rightarrow$ sum of angles = $180°$

The sum of the angles in a triangle is equal to $180°$ only in a flat, Euclidean geometry. There are curved geometries such as the surface of the Earth, where this is not the case.

A spherical triangle with one vertex at the North Pole and two vertices on the equator $90°$ apart has three right angles. Therefore the sum of the angles is equal to $270°$.

# What are similar triangles?

Two triangles are called similar if they have the same shape and differ only in size. Having the same shape means that the corresponding angles are equal. Because the sum of the angles in a triangle equals $180°$, it's sufficient that two of the angles are equal for the two triangles to be similar.

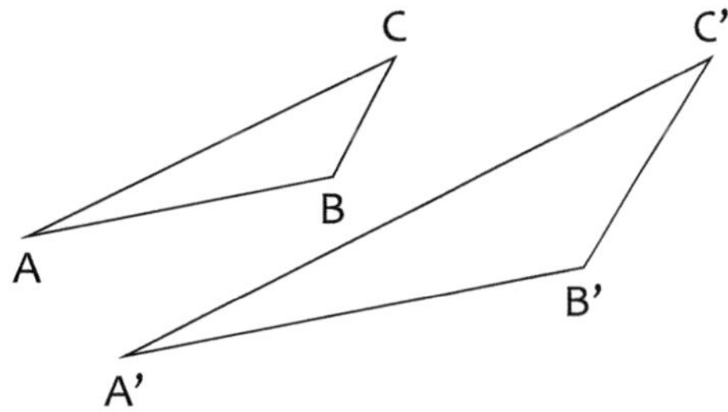

$$\Delta ABC \cong \Delta A'B'C'$$

If you place the two triangles so that the vertices $A$ and $A'$ coincide and two sides lie on top of each other, then the two other sides are parallel and you obtain the figure of the intercept theorem.

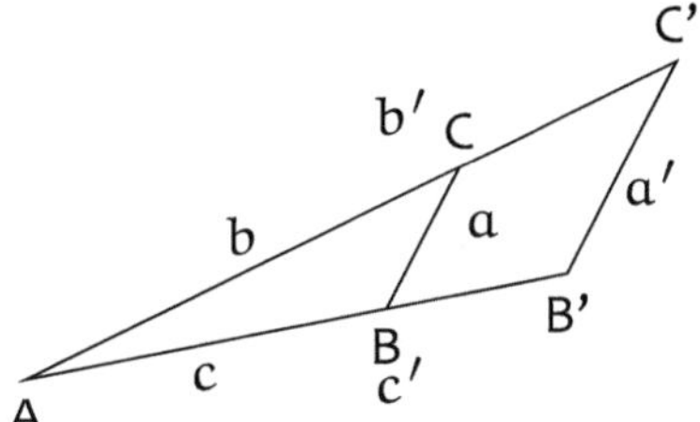

$$b = |AC|, b' = |AC'|$$
$$c = |AB|, c' = |AB'|$$

The intercept theorem states that the ratios of the corresponding lengths are equal. The ratio is the similarity factor $k$ by which one triangle must be enlarged to obtain the other.

$$\frac{a'}{a} = \frac{b'}{b} = \frac{c'}{c} = k$$

The ratio of the areas $S'$ and $S$ of the two triangles is the square of the ratio of the side lengths.

$$\frac{S'}{S} = \frac{S_{\Delta A'B'C'}}{S_{\Delta ABC}} = k^2 = \left(\frac{a'}{a}\right)^2$$

Therefore, the area of a triangle is proportional to the square of a side length, with a common proportionality factor $m$ for similar triangles.

$$S_{\Delta ABC} = ma^2$$
$$S_{\Delta A'B'C'} = ma'^2$$

# Where does the Pythagorean theorem help you?

The longest side in a right triangle (the one opposite the right angle) is called the hypotenuse. The other two sides are called legs.

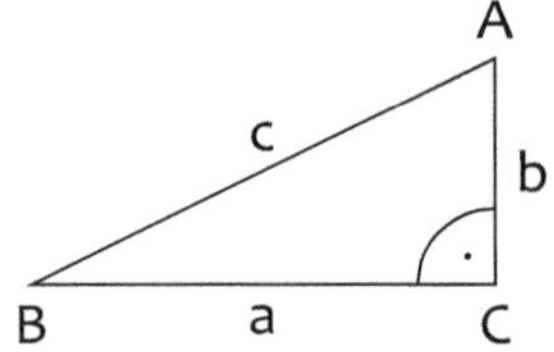

The Pythagorean theorem gives a relationship between the three side lengths of a right triangle.

$$a^2 + b^2 = c^2$$

If you are given two sides of a right triangle, you can easily calculate the third one.

$$c = \sqrt{a^2 + b^2}$$
$$a = \sqrt{c^2 - b^2}$$
$$b = \sqrt{c^2 - a^2}$$

You can prove the Pythagorean theorem by dividing the given triangle into two right triangles.

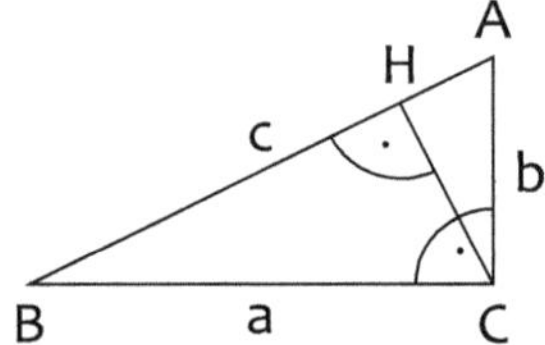

Since the corresponding angles are equal, the small triangles are similar to the original triangle.

$$\triangle ABC \cong \triangle BCH \cong \triangle CAH$$

The area of the large triangle is the sum of the areas of the two smaller triangles.

$$S_{\triangle ABC} = S_{\triangle BCH} + S_{\triangle CAH}$$

Since the three triangles are similar to each other, the area of each triangle is the same constant ($m$) times the square of its longest side, refer to the previous page.

$$mc^2 = ma^2 + mb^2$$

By dividing by the constant $m$, you obtain the Pythagorean theorem.

$$c^2 = a^2 + b^2$$

# What determines the shape of a right triangle?

If you are given an additional angle in a right triangle, you also know the remaining angle, since the sum of the angles in a triangle equals 180°.

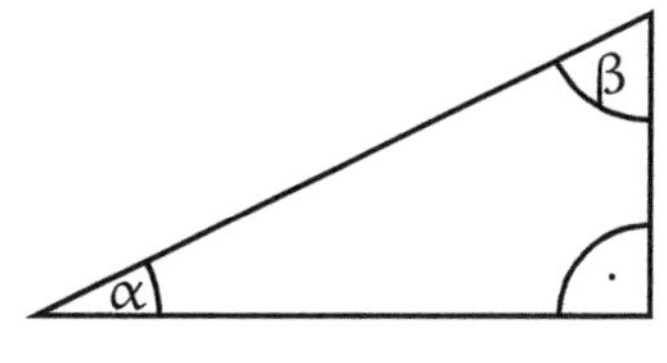

$$\alpha + \beta + 90° = 180°$$
$$\Rightarrow \beta = 90° - \alpha$$

So, if you know an additional angle in a right triangle, then you know the shape of the triangle, i.e. you know the triangle up to similarity (pure scaling by stretching).

Therefore, because of the intercept theorem, the side ratios are determined by the angle. The way the side ratios depend on the angle is described by the trigonometric functions.

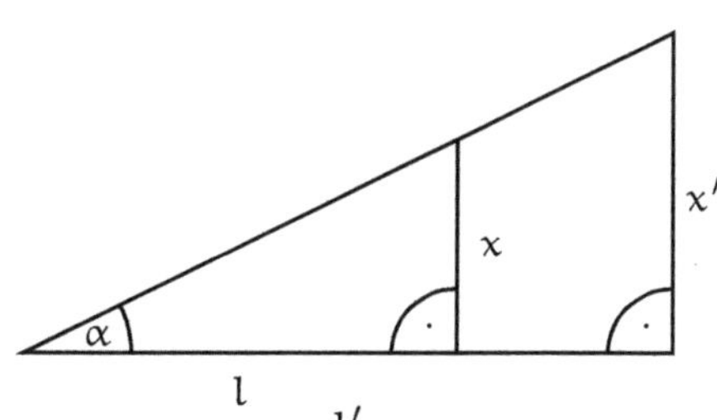

intercept theorem: $\dfrac{x}{l} = \dfrac{x'}{l'}$

# How do you calculate the side lengths of a right triangle using angles?

First, a few definitions:

The hypotenuse is the longest side in a right triangle; it's opposite the right angle. We will denote its length by "hyp" from now on.

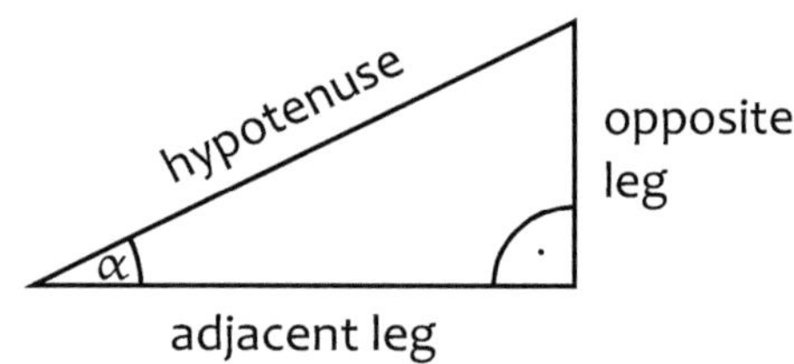


We call the given angle $\alpha$. It's the angle between the hypotenuse and the adjacent leg (with length "adj"). The leg opposite the angle $\alpha$ is called the opposite leg (with length "opp").



The trigonometric functions sine, cosine, and tangent allow you, in a right triangle, to calculate the remaining side lengths from knowing one angle and one side length.

Given $\alpha$, hyp; find opp: use sine.
Given $\alpha$, hyp; find adj: use cosine.
Given $\alpha$, adj; find opp: use tangent.

You use the same functions if the given and sought quantities are swapped.

# How is the sine defined and how do you calculate it?

The sine of the angle $\alpha$ is defined as the ratio of the opposite side to the hypotenuse.

$$\sin(\alpha) = \frac{\text{opp}}{\text{hyp}}$$

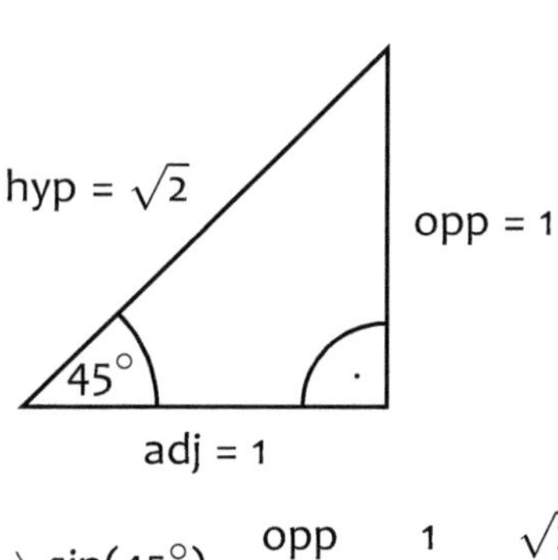

You can see that $\sin(0°) = 0$: when the angle is $0°$, the triangle collapses into a horizontal line segment, and the opposite side has length $0$.

$$\frac{\text{hyp}}{\text{adj}}$$

$$\text{opp} = 0 \Rightarrow \sin(0°) = 0$$

Similarly, $\sin(90°) = 1$: when the angle is $90°$, the triangle collapsed into a vertical line segment, and the opposite side has the same length as the hypotenuse.

$$\text{hyp} \mid \text{opp}$$

$$\text{adj} = \text{hyp} \Rightarrow \sin(90°) = 1$$

For $\alpha = 45°$ the right triangle is an isosceles triangle. If both legs have length $1$, then by the Pythagorean theorem, the hypotenuse has length $\sqrt{1^2 + 1^2} = \sqrt{2}$. Therefore, $\sin(45°) = \frac{\sqrt{2}}{2}$.

$$\Rightarrow \sin(45°) = \frac{\text{opp}}{\text{hyp}} = \frac{1}{\sqrt{2}} = \frac{\sqrt{2}}{2}$$

# How do you calculate $\sin(30°)$ and $\sin(60°)$?

For $\alpha = 30°$ you consider an equilateral triangle with side length 1, symmetrical over the horizontal. You can see that the opposite side has half the length of the hypotenuse.

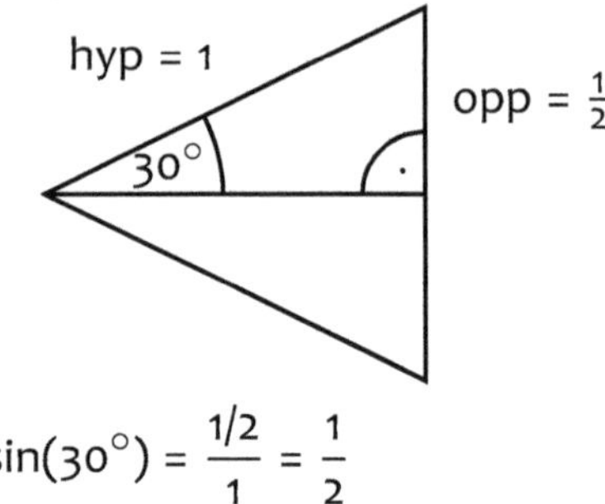


$$\Rightarrow \sin(30°) = \frac{1/2}{1} = \frac{1}{2}$$

To calculate $\sin(60°)$ you consider the angle at the top right of the triangle. The opposite side is now the horizontal segment, which according to Pythagoras has the length
$$\text{opp} = \sqrt{1^2 - (\tfrac{1}{2})^2} = \sqrt{\tfrac{3}{4}} = \tfrac{1}{2}\sqrt{3}.$$



$$\Rightarrow \sin(60°) = \frac{\text{opp}}{\text{hyp}} = \frac{\frac{1}{2}\sqrt{3}}{1} = \frac{1}{2}\sqrt{3}$$

You can easily remember these five values from the previous page and this one, as shown on the right. However, the pattern is just coincidental.

| $\alpha$ | $\sin(\alpha)$ |
|---|---|
| 0° | $\dfrac{\sqrt{0}}{2} = 0$ |
| 30° | $\dfrac{\sqrt{1}}{2} = \dfrac{1}{2}$ |
| 45° | $\dfrac{\sqrt{2}}{2}$ |
| 60° | $\dfrac{\sqrt{3}}{2}$ |
| 90° | $\dfrac{\sqrt{4}}{2} = 1$ |

# What are cosine and tangent?

The cosine of the angle $\alpha$ is defined as the ratio of the adjacent side to the hypotenuse.

$$\cos(\alpha) = \frac{adj}{hyp}$$

The tangent of the angle $\alpha$ is defined as the ratio of the opposite side to the adjacent side.

$$\tan(\alpha) = \frac{opp}{adj}$$

Thus the tangent is simply the sine divided by the cosine.

$$\tan(\alpha) = \frac{opp}{adj} =$$

$$= \frac{opp/hyp}{adj/hyp} = \frac{\sin(\alpha)}{\cos(\alpha)}$$

If you apply the Pythagorean theorem to the triangle, you find a relationship between sine and cosine. Here, as is often customary, we have written $\sin(\alpha)$ more briefly as $\sin \alpha$ and $(\sin(\alpha))^2$ as $\sin^2 \alpha$.

$$\sin^2 \alpha + \cos^2 \alpha =$$

$$= \left(\frac{opp}{hyp}\right)^2 + \left(\frac{adj}{hyp}\right)^2$$

$$= \frac{opp^2 + adj^2}{hyp^2} = \frac{hyp^2}{hyp^2} = 1$$

With this formula you can express cosine in terms of sine.

$$\cos^2 \alpha = 1 - \sin^2 \alpha$$

$$\cos \alpha = \pm\sqrt{1 - \sin^2 \alpha}$$

The sign depends on the quadrant of the angle $\alpha$.

So you can express all trigonometric functions in terms of one of them. Here the other trigonometric functions are shown in terms of sine, for the first quadrant where all trigonometric functions are positive.

$$\cos \alpha = \sqrt{1 - \sin^2 \alpha}$$

$$\tan \alpha = \frac{\sin \alpha}{\cos \alpha} = \frac{\sin \alpha}{\sqrt{1 - \sin^2 \alpha}}$$

# Where do you find sine and cosine on the unit circle?

It's helpful to remember sin, cos and tan on the unit circle, as shown on the right: in the case where the hypotenuse equals 1, sin and cos are simply the lengths of the triangle legs. If the adjacent leg equals 1, then tan is the length of the opposite leg.

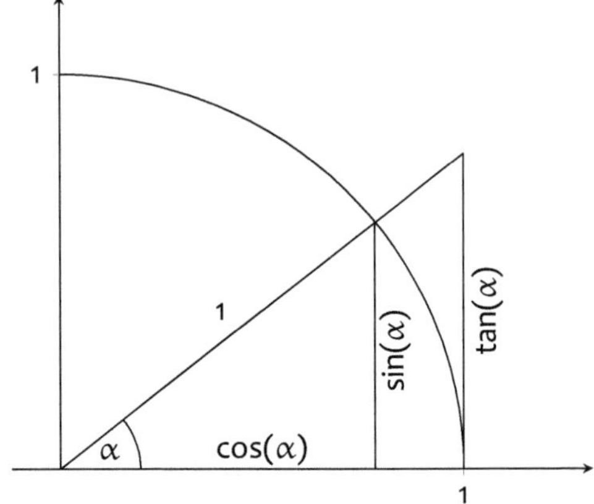



You can now imagine how these lengths change as the angle varies. You obtain the graph of the sine or cosine function by plotting the points $(\alpha, \sin(\alpha))$ or $(\alpha, \cos(\alpha))$ in a coordinate system.

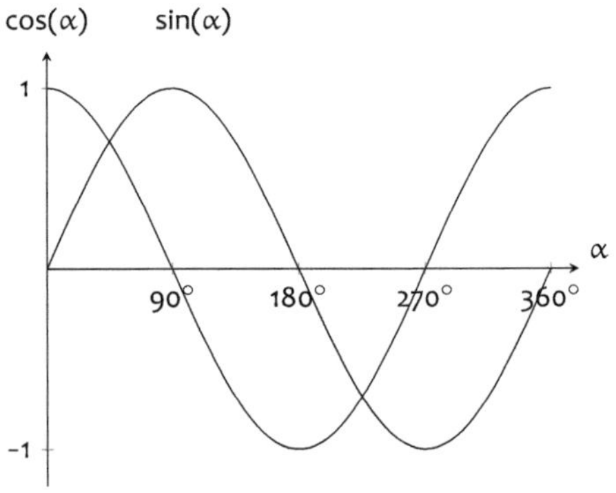

# Where do the law of cosines and the law of sines help you?

The law of cosines generalizes the Pythagorean theorem to cases where the angle is not 90°. I will omit the derivation which is possible using what you know already, because it's not essential for grasping the concept.

$$c^2 = a^2 + b^2 - 2ab\cos(\gamma)$$

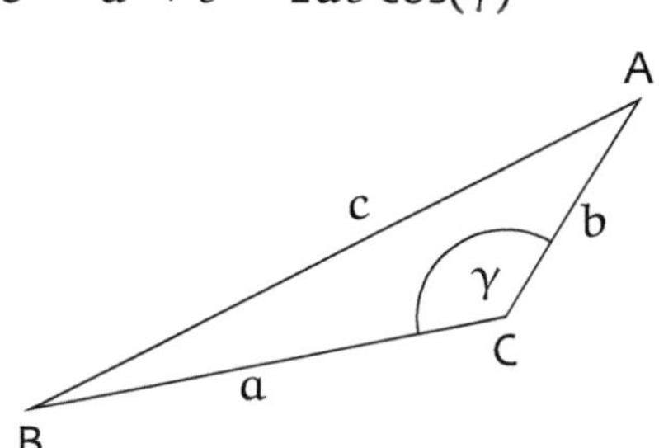

For $\gamma = 90°$ you recover the Pythagorean theorem.

$$\gamma = 90° \Rightarrow \cos(\gamma) = 0$$
$$\Rightarrow c^2 = a^2 + b^2$$

For $\gamma < 90°$ the side $c$ is shorter than in a right triangle.

$$\gamma < 90° \Rightarrow \cos(\gamma) > 0$$
$$\Rightarrow c^2 = a^2 + b^2 - 2ab\cos(\gamma) < a^2 + b^2$$

For $\gamma > 90°$ the side $c$ is longer than in a right triangle.

$$\gamma > 90° \Rightarrow \cos(\gamma) < 0$$
$$\Rightarrow c^2 = a^2 + b^2 - 2ab\cos(\gamma) > a^2 + b^2$$

For $\gamma = 180°$ the side $c$ equals the sum of $a$ and $b$.

$$\gamma = 180° \Rightarrow \cos(\gamma) = -1$$
$$\Rightarrow c^2 = a^2 + b^2 + 2ab = (a + b)^2$$

The law of sines states that the quotient of the sine of the angle and its opposite side is the same for all angles of a given triangle.

$$\frac{\sin(\alpha)}{a} = \frac{\sin(\beta)}{b} = \frac{\sin(\gamma)}{c}$$

You can easily understand this for two angle-side pairs by drawing the height to the third side.

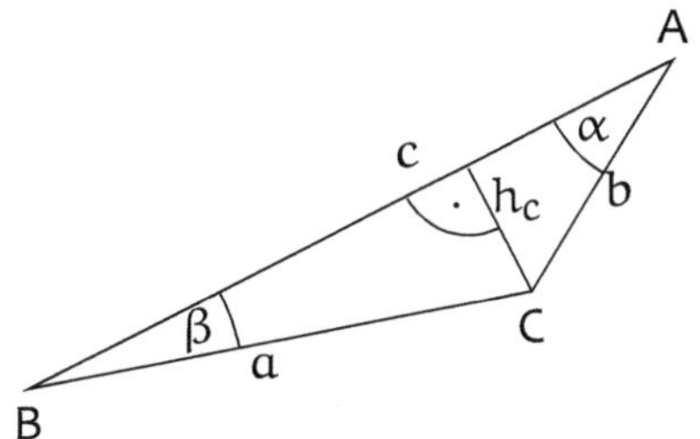

$$h_c = a\sin(\beta) = b\sin(\alpha)$$
$$\Rightarrow \frac{\sin(\alpha)}{a} = \frac{\sin(\beta)}{b}$$

# What does area have to do with multiplication?

When you multiply 3 by 5, this corresponds to 3 rows with 5 dots each, that is a total of 15 dots.

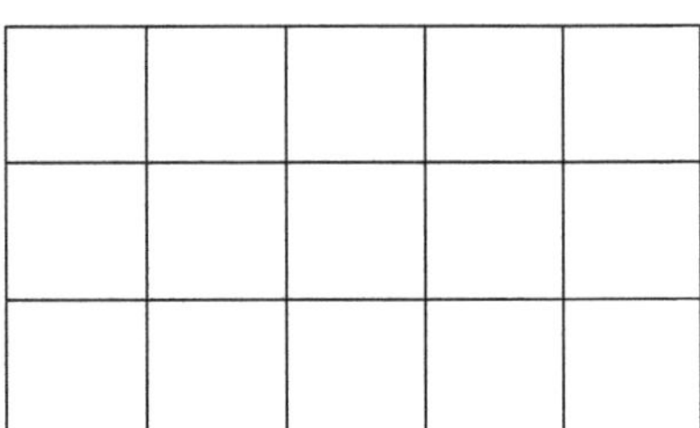

If you imagine each dot as a small square, then the total number of squares equals the number of squares along the length times the number along the width.

Area is defined as the number of unit squares. If a unit square has a length of 1 cm and therefore an area of $1\,\mathrm{cm}^2$, then a rectangle with side lengths $a$ cm and $b$ cm has an area of $(a\,\mathrm{cm}) \cdot (b\,\mathrm{cm}) = ab\,\mathrm{cm}^2$.

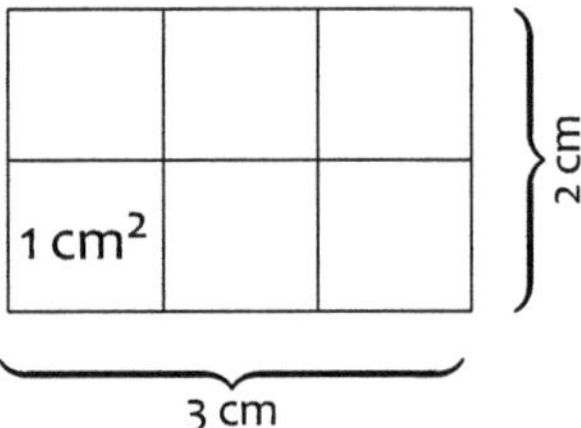



To calculate the area of another shape, you need to trace it back to the calculation of rectangles.

A triangle can be divided along the height into two triangles. This shows that the area of a triangle is half the area of the rectangle formed by the base and the height.

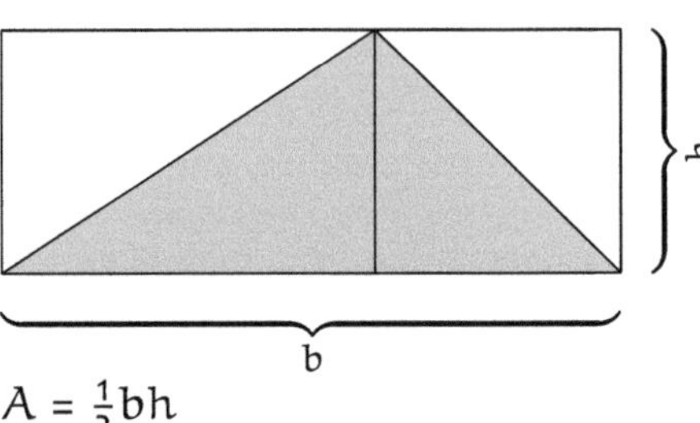



$$A = \tfrac{1}{2}bh$$

For curvilinearly bounded areas, you generally need infinitely many "infinitely small rectangles". This leads to integral calculus.

# How can you remember the binomial formulas using areas?

If you interpret the square in the binomial formula $(a + b)^2 = a^2 + 2ab + b^2$ as an area, you can visualize the formula. You obtain the large square $(a + b)^2$ by adding to the square $a^2$ two rectangles $ab$ and the small square $b^2$.

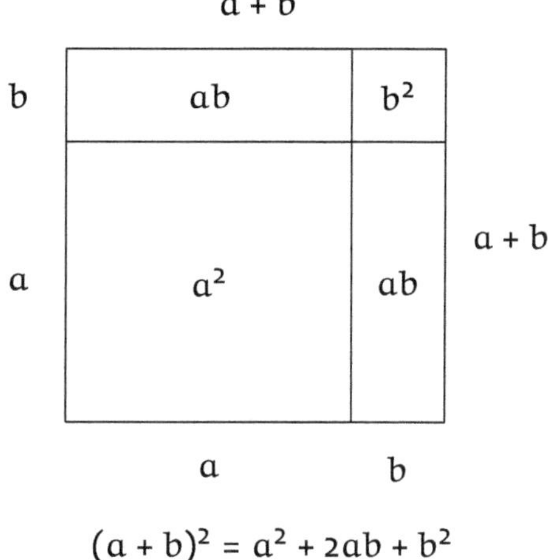


$$(a + b)^2 = a^2 + 2ab + b^2$$

For a visual understanding of the formula $(a - b)^2 = a^2 - 2ab + b^2$ you can see in the figure that from the large square you first have to subtract twice the area of the rectangle with side lengths $a$ and $b$ and the area $ab$ (the two rectangles with thick borders).

However, in doing so, you have subtracted the small square $b^2$ twice, so you have to add it back once.

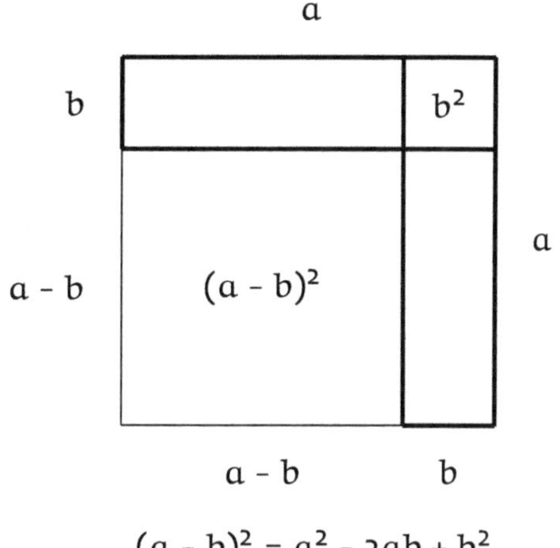


$$(a - b)^2 = a^2 - 2ab + b^2$$

# How do you calculate a volume?

For volume it's just like with area calculation, except that you multiply three numbers together: width, length, and height.

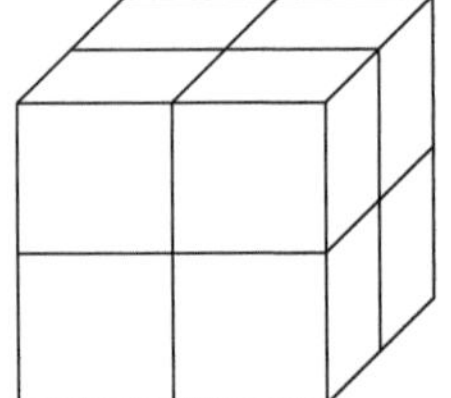

$$V = 2 \cdot 2 \cdot 2 = 2^3$$
$$V = a \cdot a \cdot a = a^3$$

For a prism or a cylinder, the volume is simply the base area times the height. You can understand this by imagining the solid divided into small rods, the volume of each of it equals its base area times its height.

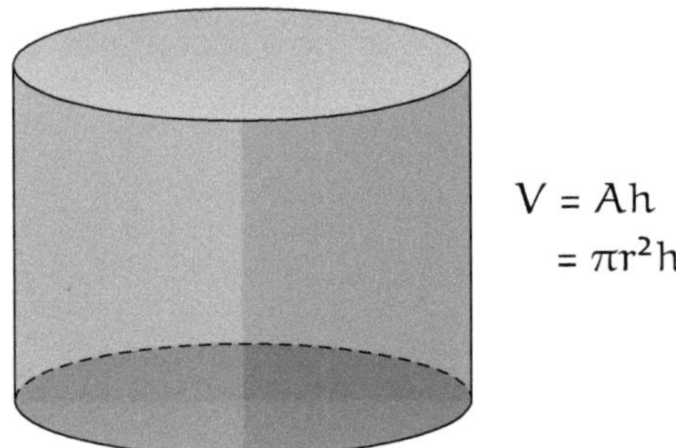

$$V = Ah$$
$$= \pi r^2 h$$

For a pyramid and a cone the volume is one third of the base area times the height. This is similar to the formula for the area of a triangle which is one half of base times height.

The reason why you have to use "one third" is that volume is three-dimensional. You can understand this by integration or by dividing a cuboid into three pyramids of equal size.

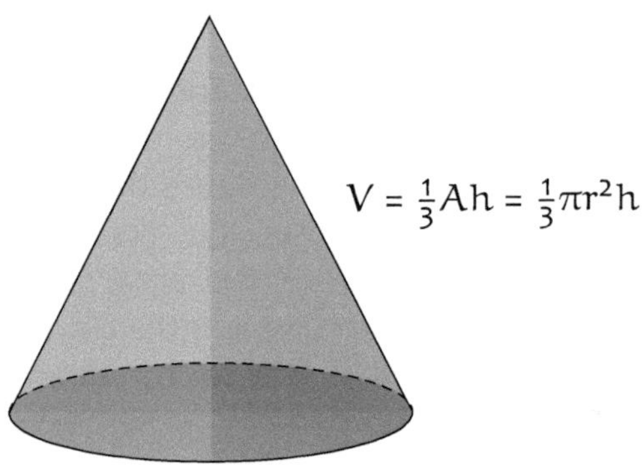

$$V = \tfrac{1}{3}Ah = \tfrac{1}{3}\pi r^2 h$$

The volume of a sphere is similarly given as one third of the surface area times the radius. I will not explain why the surface area of a sphere is $4\pi r^2$.

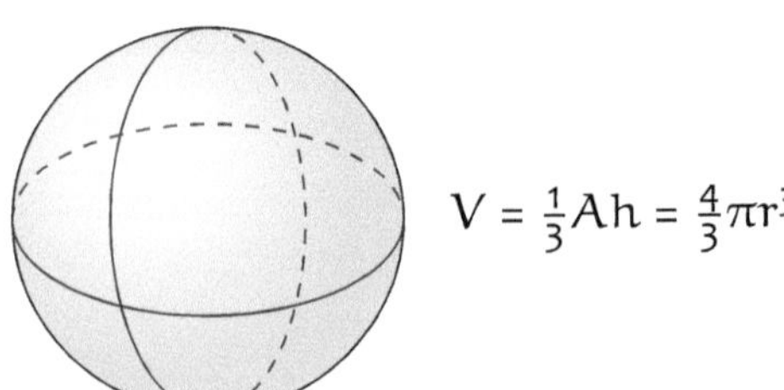

$$V = \tfrac{1}{3}Ah = \tfrac{4}{3}\pi r^3$$

# What do exponents have to do with dimensions?

You start with a point.

If you move the point by the distance $a$, you get a segment of length $a$.

If you shift the segment of length $a$ perpendicularly by $a$, you get a square of area $a^2$. That is why "raising to the second power" is also called squaring.

If you shift the square of side length $a$ perpendicularly by $a$, you get a cube of volume $a^3$.

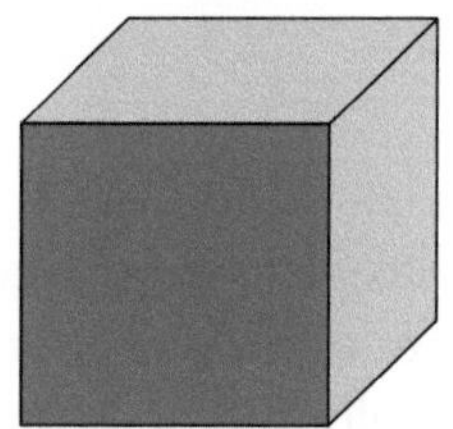

The corresponding unit of measure also carries the power. In this way, you can continue in an abstract way ($cm^d$). The exponent is the dimension $d$.

unit measure of length: 1 cm
unit measure of area: 1 cm$^2$
unit measure of volume: 1 cm$^3$

If you scale a length, area, or volume by a factor of $s$, the corresponding quantity grows by a factor of $s^d$, where the dimension $d = 1, 2, 3$. This applies not only to squares and cubes, but to any area and any volume.

length scaling: $(sa)^1 = s \cdot a$
area scaling: $(sa)^2 = s^2 \cdot a^2$
volume scaling: $(sa)^3 = s^3 \cdot a^3$

As you can see, the area grows with the second power, the volume with the third. Therefore, the surface area grows more slowly than the volume.

A bear has a better ratio of body volume (heat production) to skin surface area (heat loss) than a mouse, which is why there are polar bears but no polar mice.

# What is $\pi$ and why do you need this number?

The circumference of a circle is proportional to its radius. If you double the radius, the circumference also doubles.

$$C \propto r$$

The proportionality constant is $2\pi$. This can be used as a definition for $\pi$.

$$C = 2\pi r$$

One gets $\pi = 3.141592653\ldots$

There are many series representations for $\pi$, see e.g. on the right side one by Madhava (14th century) and Leibniz (1682), and one by Ramanujan (1914).

$$\frac{\pi}{4} = \sum_{n=1}^{\infty} \frac{(-1)^n}{2n+1} = 1 - \frac{1}{3} + \frac{1}{5} - \frac{1}{7} + \ldots$$

$$\frac{1}{\pi} = \frac{2\sqrt{2}}{9801} \sum_{n=0}^{\infty} \frac{(4n)!(1103 + 26390n)}{(n!)^4 396^{4n}}$$

There are also easy-to-remember and very accurate approximation formulas for $\pi$, usually the first one is sufficient.

$$\pi \approx 3.14$$

$$\pi \approx \frac{9}{5} + \sqrt{\frac{9}{5}} = 3.141640786\ldots$$

$$\pi \approx \frac{355}{113} = 3.141592920\ldots$$

$\pi$ is needed when calculating the area of a circle. You will understand why this formula holds on the next pages.

$$A = \pi r^2$$

The surface area of a sphere also contains $\pi$, as does the volume of the sphere.

$$S = 4\pi r^2$$

$$V = \frac{4}{3}\pi r^3$$

$\pi$ appears not only in geometry, but also in many other areas of mathematics.

Gaussian normal distribution:

$$\varphi(x) = \frac{1}{\sqrt{2\pi}} e^{-\frac{x^2}{2}}$$

# What is the radian measure and why is this unit of measure practical?

If you have a segment of the circumference of a circle with angle $\alpha$, the arc length is just the pro-rata part of the total circumference.



$$C(\alpha) = \frac{\alpha}{360°} \cdot C(360°) =$$

$$= \frac{\alpha}{360°} \cdot 2\pi r = \alpha \frac{\pi}{180°} r$$

The angle, for which the arc length of the unit circle ($r = 1$) segment is equal to 1, is called 1 radian.

$$1\,\text{rad} = \frac{180°}{\pi}$$

The arc length is then simply the angle in radians multiplied by the radius.

$$C = \alpha\, r$$
$$\alpha \text{ in rad}$$

The specification of angles in radians is common and especially very practical in calculus.

| $30°$ | $45°$ | $60°$ | $90°$ | $180°$ | $360°$ |
|---|---|---|---|---|---|
| $\dfrac{\pi}{6}$ | $\dfrac{\pi}{4}$ | $\dfrac{\pi}{3}$ | $\dfrac{\pi}{2}$ | $\pi$ | $2\pi$ |

When you give an angle in radians, you don't use the degree symbol. The unit of measure "rad" is rarely written.

$$\alpha = \frac{\pi}{2}, \qquad \beta = \pi$$

$$\sin\left(\frac{\pi}{2}\right) = 1, \qquad \sin(\pi) = 0$$

# How do you calculate the area of a circle?

The area A of a circle is proportional to the square of the radius. The proportionality constant is $\pi$.

$A = \pi r^2$

To understand this, we divide the area of the circle into $n$ pie slices, shown on the right for $n = 20$.

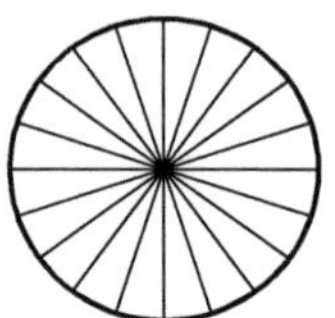

If you arrange them alternately pointing up and down, a rectangle-like area is formed.

$\approx \pi r$

$\approx r$

This holds because the sum of the upper and lower edges is equal to the circumference.

Its length is approximately half the circumference of the circle ($2\pi r/2 = \pi r$), and its height is approximately the radius ($r$).

For $n \rightarrow \infty$, the deviation becomes arbitrarily small. Therefore, the area of the circle is equal to $\pi r^2$.

For $n \rightarrow \infty$ it follows
$A = \pi r \cdot r = \pi r^2$.

# Why do you need a coordinate system?

With a coordinate system, you can assign two real numbers as coordinates to points on the plane in a one-to-one way, and vice versa. This creates very useful bridges between geometry, algebra and analysis.

Most simply and most commonly, the Cartesian coordinate system is used, consisting of two number lines perpendicular to each other.

The $x$-value indicates how far to the right ($x > 0$) or left ($x < 0$) a point sits; the $y$-value indicates how far up ($y > 0$) or down ($y < 0$) a point sits.

The $x$-axis consists of the points that are centered with respect to up/down, that is to say of all points with $y = 0$.

The $y$-axis consists of the points that are centered with respect to right/left, that is to say of all points with $x = 0$.

The point that is centered with respect to both dimensions is the intersection of the $x$- and $y$-axes, so its coordinates are $y = x = 0$.

The distance of a point from the origin can be calculated using the Pythagorean theorem, as can the distance between two points.

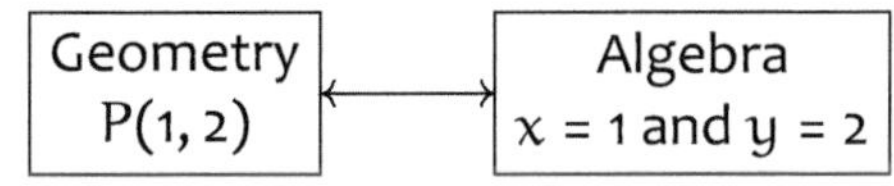



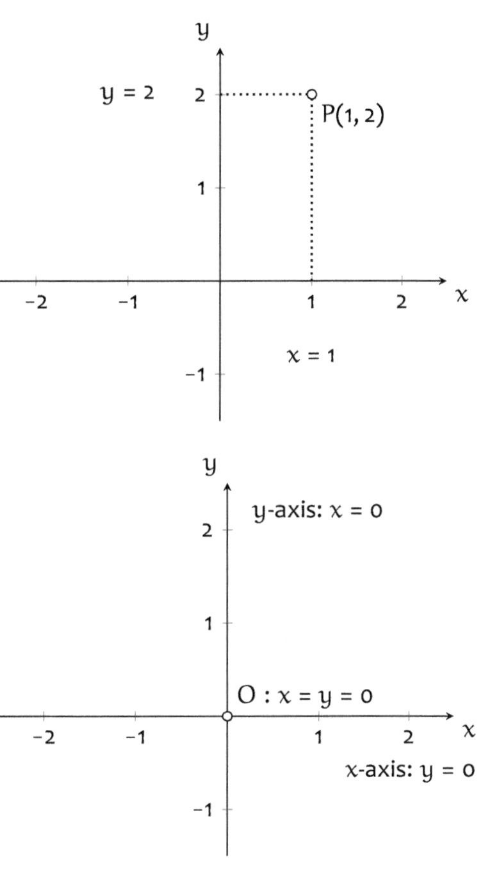



$$|OP| = \sqrt{x^2 + y^2}$$

$$|P_1P_2| = \sqrt{(x_2 - x_1)^2 + (y_2 - y_1)^2}$$

# Solving Word and Application Problems

Mathematics has an incredible number of applications.

Mathematics is found in nature and technology. With mathematics, you can predict natural events such as solar eclipses (very precisely) or volcanic eruptions (only with probabilities).

With mathematics, other sciences can tame nature and ultimately build smartphones with navigation and translation apps from sand, oil, and a few other ingredients.

Word problems are simple applications.

You solve them by translating them into mathematics, solving the problem within mathematics, and then translating the solution back.

Translating a real-world problem into mathematics usually requires knowledge from other sciences. However, the following examples are so simple that everyday knowledge is sufficient.



A. Gründers, *Math Made Clear: From the Basics to Calculus*, https://doi.org/10.1007/978-3-662-73221-2_6

# What does math have to do with reality?

There is no consensus on whether mathematical objects exist independently from humans and are merely discovered, or whether they are invented by people.

Mathematics:
Discovered or invented?

Either way, it is surprising how well mathematics can quantitatively describe and predict certain aspects of what we call reality.

planetary motion (Kepler, 1619)
electrodynamics (Maxwell, 1865)
theory of relativity (Einstein, 1916)
quantum theory (Heisenberg, 1925)
Only one scientist is named in each case.

With mathematics you can calculate some properties of elementary particles with an amazing degree of accuracy with experiments. The calculation involves more than 10 000 multiple integrals from Feynman diagrams.

magnetic moment of the electron as dimensionless g-factor:
- theoretical: $g = 2.002\,319\,304\ldots$
- experimental: $g = 2.002\,319\,304\ldots$

Mathematics in combination with other sciences is the foundation for many scientific discoveries and technological achievements.

mathematics
+ thermodynamics: gasoline engine
+ electrodynamics: electric motor
+ semiconductor theory: laptop
+ computer sciences: AI

# How do you solve word problems?

Word problems refer to reality. To solve a word problem, you first need to translate it into a mathematical model. This means that you have formulated the question mathematically.

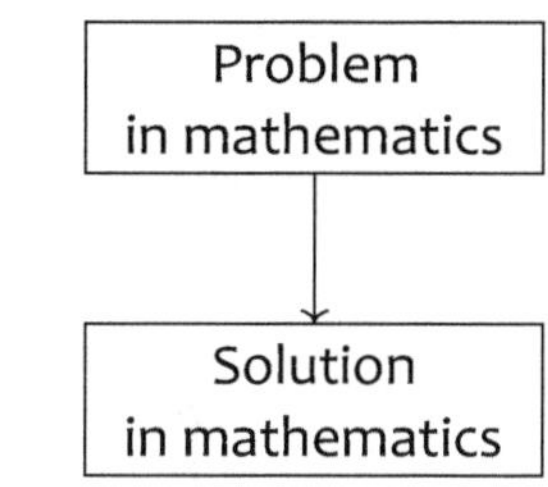


Secondly, you solve the mathematical problem.

Thirdly, you translate the solution from the mathematical model back into reality.



If you have done everything correctly you get a modeling square.

The upward pointing arrow on the left arises because the "solution in reality" can lead to a more precise or extended problem description. With this new problem, the square can be run through again.

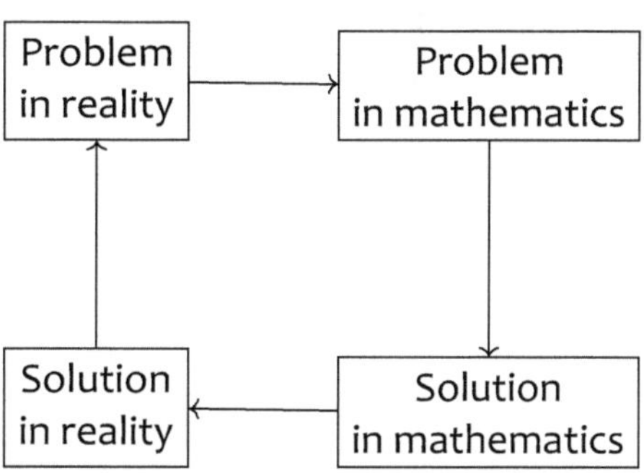

# How do you translate a word problem into mathematics?

To translate the word problem into mathematics:

- Read the text very carefully.

- For each sentence, see what information you get, and write it down mathematically.

- Denote known quantities with letters, such as $a, b, \ldots$ or e.g. $e$ for "eggs" and state their value. Write them under "Given".

- Understand what the unknown quantities are and denote them with letters as well. Write them under "Wanted".

- Establish a relationship between the variables. Often it's helpful to make a sketch.

One and a half chickens lay one and a half eggs in one and a half days. How many eggs does one chicken lay in one day?

Given:
number of chickens (1st phrase): 1.5
number of days (1st phrase): 1.5
number of eggs (1st phrase): 1.5
number of chickens (2nd phrase): 1
number of days (2nd phrase): 1

Given:
$ch_1 = 1.5$, $d_1 = 1.5$, $e_1 = 1.5$
$ch_2 = 1$, $d_2 = 1$

Wanted:
number of eggs $e_2$

Idea: The egg-laying rate per day and chicken is constant.
Therefore: egg-laying rate =

$$= \frac{e_1}{ch_1 \cdot d_1} = \frac{e_2}{ch_2 \cdot d_2}.$$

# How do you solve the problem in mathematics?

To solve the problem in mathematics:

- Check whether you have found all the necessary information about the given situation and the unknowns from the text.

(continuation of the example)

Equation: $\dfrac{e_1}{ch_1 \cdot d_1} = \dfrac{e_2}{ch_2 \cdot d_2}$

Given:

$ch_1 = 1.5,\ d_1 = 1.5,\ e_1 = 1.5$

$ch_2 = 1,\ d_2 = 1$

Wanted: $e_2$

- You know it once the problem also makes sense mathematically, i.e. it is solvable.

Ok, since you have one equation for one unknown.

- Consider what kind of mathematical problem it is. Do you need to solve equations or determine special values, such as a minimum?

Solving an equation

- Often, further implicit conditions arise from the problem statement. Write these down as well, using the variable names you introduced.

The number of eggs should be positive ($e_2 > 0$).
It does not have to be a natural number, as it wasn't in the question either, since these are average values.

- Consider which methods you know to solve the equation.

Solving the equation for $e_2$ by multiplying it by $ch_2 \cdot d_2$

- Solve the equation. Check the solution using the original equation. Substitute the solution into the original equation in order to double-check if it makes sense.

$$e_2 = ch_2 \cdot d_2 \cdot \dfrac{e_1}{ch_1 \cdot d_1} =$$

$$= 1 \cdot 1 \cdot \dfrac{1.5}{1.5 \cdot 1.5} = \dfrac{1}{1.5} = \dfrac{1}{\frac{3}{2}} = \dfrac{2}{3}$$

The solution satisfies the original equation and is positive.

# How do you translate the solution back into reality?

To bring the solution from mathematics back into reality:

■ Formulate the solution as a sentence using the words from the problem statement.

■ Consider, whether the solution makes sense to you, whether it seems intuitively plausible.

■ State, if you can think of anything else that can be concluded from the solution.

■ If the result is not satisfactory, formulate a refined question.

(continuation of the example)

Based on the information given in the problem statement, it follows that on average, a chicken lays 2/3 eggs per day.

Yes, the solution makes sense: from the problem statement, it quickly follows that a chicken lays one egg in one and a half days, so in one day it lays less than one egg.

The average number of eggs that $ch$ chickens lay in $d$ days is $e = \frac{2}{3} \cdot ch \cdot d$.

If a chicken lays an average of 2/3 eggs per day: On what proportion of days does the chicken lay an egg, and on what proportion does it not? How does this change if it lays 2 eggs on a certain proportion of days?

# What is the rule of three useful for?

It often happens, that if you double one quantity, another quantity also doubles.

The double quantity costs twice as much.
A doubled-sized square has twice the perimeter.

If this holds not only for the factor 2, but for any factor, the quantities are called proportional to each other.

The price is proportional to the quantity.
The perimeter is proportional to the side length.

We write a squashed $\alpha$ for it, "$\propto$", in some regions a simple wavy line "$\sim$". Please do not confuse these symbols with the double wavy line "$\approx$" for approximately.

$l \propto a, \; l \sim a$   (proportional to )
$\frac{1}{3} \approx 0.33$   (approximately equal)

When there are two corresponding quantities, sometimes the correspondence sign "$\cong$" is used, which in geometry also means congruent.

2 kg corresponds to 3 €.
2 kg $\cong$ 3 €

If you have two quantities that are proportional to each other, and for the first quantity the corresponding second one is given, you can easily calculate the second quantity even if the first quantity has a different value.

2 kg of potatoes costs 3 €.
How much does 3 kg of potatoes cost?

You write down first what you know. Second, you convert to a unit quantity. As a third step (rule of three), you multiply in order to get the desired amount.

2 kg $\cong$ 3 €
1 kg $\cong$ 1.50 €
3 kg $\cong$ 4.50 €

Of course, you could also simply multiply by the factor by which the second value is larger (or smaller).

$$3 \text{ kg} \cong \frac{3\text{kg}}{2\text{kg}} \cdot 3\,€ = 1.5 \cdot 3\,€ = 4.50\,€$$

# How do percentage and interest calculations work?

Fractions are often given as a percentage. You can simply calculate with a percentage as with a number, where the percent sign stands for $1/100 = 0.01$.

$$\% = 1\% = 0.01$$
$$20\% = 20 \cdot 0.01 = 0.2$$

If you want to write a number as a percentage, simply multiply it by the number 1 (which does not change anything), which you write as $1 = 100\%$.

$$0.2 = 0.2 \cdot \underbrace{100\%}_{=1} = \underbrace{0.2 \cdot 100}_{=20}\% = 20\%$$

With per mille it's exactly the same, only that one per mille, 1‰, stands for $1/1000 = 0.001$.

$$0.035 = 0.035 \cdot \underbrace{1000‰}_{=1} =$$
$$= \underbrace{0.035 \cdot 1000}_{=35}‰ = 35‰$$

If you start from a base value, then the portion that corresponds to a percentage rate is called the percentage value.

base value: 100 € (store price)
percentage rate: 19% (tax rate)
percentage value: 19 € (tax)

The percentage value has the same unit as the base value, and their ratio is the percentage rate.

$$19 \,€ \div 100 \,€ = \frac{19}{100} = 0.19 = 19\%$$

In interest calculations the base value is the initial capital, the percentage rate is the interest rate, and the percentage value is the interest.

base value: 100 € (capital)
percentage rate: 3% (interest rate p.a.)
percentage value: 3 € (interest after 1y)

If the capital is invested for several years, the interest is also compounded, i.e. in the second year, both the capital and the interest earn interest.

capital at the beginning: 100 €
capital after 1 year:
$$100 \,€ + 3\% \cdot 100 \,€ = 1.03 \cdot 100 \,€$$
capital after 2 years $= 1.03^2 \cdot 100 \,€$

# How does compound interest work?

Interest means, that you receive a portion (according to the interest rate) in addition to the loaned capital when you lend someone a certain amount of capital.

With compound interest you also lend the interest to the debtor, so in the second year you also earn interest on the interest, and so on.

If you want to calculate how long it takes for your capital to double at a given interest rate, you can find this out by taking logarithms.

Here you can see the number of years it takes to double for some interest rates. For small percentages, the duration is approximately 69.3 (= $100 \cdot \log(2)$) years divided by the percentage rate.

capital C, e.g. 100 € (capital)
percentage rate: p, e.g. 3% = 0.03
capital in year 1: $(1 + p)C$, e.g. 103 €

capital after year 1: $(1 + p)C$
capital after year 2: $(1 + p)^2 C$
...
capital after year n: $(1 + p)^n C$

doubling: $(1 + p)^n C = 2C \quad | \div C$

$(1 + p)^n = 2 \quad | \log$

$n \log(1 + p) = \log(2) \quad | \div \log(1 + p)$

$$n = \frac{\log(2)}{\log(1 + p)}$$

| p | 1% | 2% | 3% | 5% | 10% |
|---|---|---|---|---|---|
| n | 69.7 | 35.0 | 23.4 | 14.2 | 7.3 |

# Why are units of measure important?

So far, we have been calculating with numbers.

Geometric and physical quantities are not just pure numbers, but have a unit of measure, also called unit of measurement.

A table is not 1 long,
it is 1 meter long.

A unit of measure is a predefined size of a quantity that is standardized.

meter, second, kilogram
abbreviated: m, s, kg

Lengths, areas, masses, etc. are given as multiples of the unit of measure. The multiplication dot is usually omitted.

$5 \text{ meters} = 5 \cdot 1 \text{ meter} = 5 \text{ m}$

You can convert units of measure.

$1 \text{ m} = 100 \text{ cm}$

When converting powers of units of measures like square meters or cubic meters, you must also raise the factor to the corresponding power.

$1 \text{ m}^2 = (1 \text{ m})^2 = (100 \text{ cm})^2$
$= (100)^2 \text{ cm}^2 = 10000 \text{ cm}^2$

You can treat units of measure in calculations like factors.

How much is 1 km/h in m/s?

$$1 \frac{\text{km}}{\text{h}} = 1 \frac{1000 \text{ m}}{60 \cdot 60 \text{ s}} =$$

$$= \frac{5}{18} \frac{\text{m}}{\text{s}} \approx 0.278 \frac{\text{m}}{\text{s}}$$

Percentage is, in this sense, a kind of unitless unit of measure, as is radian (which is also often omitted) and the degree symbol.

$1 \% = 0.01$

$$1 \text{ rad} = \frac{180^\circ}{\pi} \approx 57.3^\circ$$

# How can you check or derive formulas using units of measure?

If you have an applied formula, you can use the units of measure to check whether your equation can be correct. The result should be in the unit of measure you expect; this is a good way to verify your work.

Time for free fall $t = \sqrt{\frac{2h}{g}}$,

for $h = 10\,\text{m}$ and $g = 9.81\frac{\text{m}}{\text{s}^2}$:

$$t = \sqrt{\frac{20\,\text{m}}{9.81\frac{\text{m}}{\text{s}^2}}} \approx \sqrt{2.04\frac{\cancel{\text{m}}\,\text{s}^2}{\cancel{\text{m}}}} = 1.43\,\text{s}$$

If you want to determine the formula for the period T of a pendulum on a planet, then the problem only involves three quantities: the length of the pendulum L, the mass M that you hang, and the weight $W$ of the mass.

$$[T] = \text{s}$$
$$[L] = \text{m}$$
$$[M] = \text{kg}$$
$$[W] = \text{N} = \frac{\text{m}\,\text{kg}}{\text{s}^2}$$

You make the assumption that the period is a product of powers of length, mass, and weight.

$$T \propto L^a M^b W^c$$

$\Rightarrow$ for the units:
$$\text{s} = \text{m}^a \cdot \text{kg}^b \cdot \left(\frac{\text{m}\,\text{kg}}{\text{s}^2}\right)^c$$

If you compare the exponents of the units on the left and right, you get three equations.

Exponent of s: $1 = -2c$
Exponent of m: $0 = a + c$
Exponent of kg: $0 = b + c$

You can easily solve the three equations, starting with the first one.

$$c = -\tfrac{1}{2},\ a = \tfrac{1}{2},\ b = \tfrac{1}{2}$$

Thus, you obtain an equation for the period.

$$T \propto \sqrt{\frac{L \cdot M}{G}}$$

If you also know that weight and mass are proportional to each other, you get the dependence of the period on the length and gravitational acceleration.

$$W = mg,\ g = 9.81\frac{\text{m}}{\text{s}^2}$$

$$T \propto \sqrt{\frac{l}{g}}$$

Only the correct constant, here $2\pi$, cannot be determined this way.

$$T = 2\pi\sqrt{\frac{l}{g}}$$

# Functions

When one quantity depends on another, this is described using functions.

A function f uniquely maps one number, say $x$, to another number, denoted $f(x)$. It assigns to a value of the independent variable a value of the dependent variable.

The use of functions and the study of their properties is essential in mathematics and its applications.

This allows you to formulate and answer many questions, e.g. how one quantity depends indirectly on another, or when a quantity becomes maximal as a function of another.

In some places I use limits and derivatives, which will be explained in detail later. You can skip these parts if you are not familiar with these concepts.

If the range of values consists of mathematical objects other than numbers, the term "map" is used instead of "function".

Addition and multiplication are also functions; the expression $a + b$ is the shorthand for a function of two variables add$(a, b)$.



A. Gründers, *Math Made Clear: From the Basics to Calculus*,
https://doi.org/10.1007/978-3-662-73221-2_7

# What is a function?

Often, one variable depends on another variable. The independent variable is often called $x$ and the dependent variable $y$.

One writes $y = f(x)$ and calls $f$ a function. The parentheses indicate that you are taking the function $f$ of $x$.

The result of applying the function $f$ to a specific value $x$ is denoted as $f(x)$.

Sometimes the parentheses are omitted. You should generally not do this, except in special cases.

Some mathematical symbols have a "built-in" parenthesis, such as the square root sign or the exponent.

A function is an assignment from a domain (here $\mathbb{R}$) to a codomain (here also $\mathbb{R}$). Each element of the domain is assigned to exactly one element of the codomain. Not all elements of the codomain need to be hit by the function.

The domain is the set for which you want to and can define the function. For example, the zeros of the denominator must be excluded.

You can also write the domain after the function, either as a set or as a condition.

$x$: (often) independent variable
$y$: (often) dependent variable

$y = f(x)$
$f$: function
$x$: argument, in parentheses

$f(x)$: result of applying $f$ to $x$
$$f(x) = x^2 \Rightarrow f(3) = 9$$

$f(x) = \sin(x)$, sometimes also $\sin x$
$g(x) = \sin(2x)$, sometimes also $\sin 2x$

$$f(x) = \sqrt{x + 3}$$
$$g(x) = 2^{x+3}$$

$$f : \mathbb{R} \to \mathbb{R}$$
$$x \mapsto f(x) = x^2$$

The elements of the codomain that are hit by the function form the image (or the range) of the function.

$$f : \mathbb{R} \setminus \{0\} \to \mathbb{R}$$
$$x \mapsto \frac{1}{x}$$

$$f(x) = \frac{1}{x}, x \in \mathbb{R} \setminus \{0\}$$

$$f(x) = \frac{1}{x}, x \in \mathbb{R}, x \neq 0$$

# How can you represent a function?

A function, i.e. a dependency of a variable $y$ on an independent variable, can be represented in different ways.

For example, you can specify the function rule, i.e. a rule by which to calculate $f(x)$ from $x$, and the domain of $x$.

If $x$ can only take a few values or you only want to display a few values, you can use a table.

If $x$ is a real number, a graphical representation is suitable. For each $x$-value, you plot a point $P(x, y)$ with the corresponding $x$- and $y$-values in a coordinate system.

If $x$ runs through the real numbers, you get many points, which can, for example, lie on a line. In this way, the graph of the function emerges.

$$y = f(x) = x^2, \quad x \in \mathbb{R}$$

| $x$ | 0 | 1 | 2 | 3 | 4 |
|---|---|---|---|---|---|
| $f(x) = x^2$ | 0 | 1 | 4 | 9 | 16 |

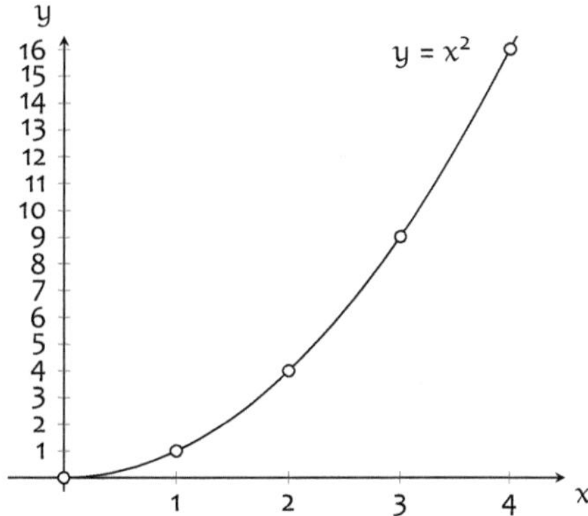


It's helpful if you can easily switch between the function rule and the graphical representation.

# What does "injective" and "surjective" mean?

A function assigns to each element of the domain (independent variable) exactly one element of the codomain (dependent variable).

An assignment that assigns to one element of the domain two elements of the codomain does not define a function.

The terms "injective" and "surjective" characterize a function based on its codomain.

A function $f : D \to E, x \mapsto f(x)$ is called injective if for every $y \in E$ there is at most one $x$ with $y = f(x)$.

If there is a $y \in E$ that is the function value of two different $x \in D$, then the function is not injective.

A function $f : D \to E, x \mapsto f(x)$ is called surjective if for every $y \in E$ there is at least one $x$ with $y = f(x)$. So every element of the codomain is hit by the function. So, for a surjective function the range (or image) is the same as the codomain.

If there is a $y \in E$ for which no $x \in D$ can be found with $y = f(x)$, then the function is not surjective.

$$f : \mathbb{R} \to \mathbb{R}, x \mapsto f(x) = x^2$$
$$2 \mapsto 4$$
f is a function.

$$f : \mathbb{R} \to \mathbb{R}, x \mapsto y \text{ with } y^2 = x$$
$$4 \mapsto \pm 2$$
f is not a function.

What matters is the number of $x$-values that correspond to a given $y$-value: $|\{x \in D \mid f(x) = y\}|$.

$$|\{x \in D \mid f(x) = y\}| \leq 1 \text{ for all } y \in E$$
$$f : \mathbb{R}_{\geq 0} \to \mathbb{R}, x \mapsto y = f(x) = x^2$$
$$|\{x \in \mathbb{R}_{\geq 0} \mid x^2 = 4\}| = |\{2\}| = 1 \leq 1$$
f is injective.

For injective functions, every line parallel to the x-axis intersects the graph of the function at most once.

$$f : \mathbb{R} \to \mathbb{R}, x \mapsto y = f(x) = x^2$$
$$|\{x \in \mathbb{R} \mid x^2 = 9\}| = |\{\pm 3\}| = 2 \nleq 1$$
f is not injective.

$$|\{x \in D \mid f(x) = y\}| \geq 1 \text{ for all } y \in E$$
$$f : \mathbb{R} \to \mathbb{R}_{\geq 0}, x \mapsto y = f(x) = x^2$$
$$|\{x \in \mathbb{R} \mid x^2 = 9\}| = |\{3\}| = 1 \geq 1$$
f is surjective.

For surjective functions, every line parallel to the x-axis in the codomain intersects the graph of the function at least once.

$$f : \mathbb{R} \to \mathbb{R}, x \mapsto y = f(x) = x^2$$
$$|\{x \in \mathbb{R} \mid x^2 = -1\}| = |\{\}| = 0 \ngeq 1$$
f is not surjective.

# What does "bijective" mean?

A function $D \to E, x \mapsto y = f(x)$ is called bijective if for every $y \in E$ there is exactly one $x$ with $y = f(x)$.

$|\{x \in D \mid f(x) = y\}| = 1$ for all $y \in E$

A function is therefore bijective if and only if it is injective and surjective.

(injective: $|\{x \in D \mid f(x) = y\}| \leq 1$ and surjective: $|\{x \in D \mid f(x) = y\}| \geq 1$)
$\Leftrightarrow$ bijective: $|\{x \in D \mid f(x) = y\}| = 1$

Each x-value corresponds to exactly one y-value and each y-value corresponds to exactly one x-value.

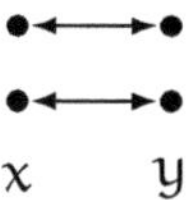


Instead of "bijective", we can also say "one-to-one".

If a function is bijective, it can be inverted: you can define an inverse function that assigns to each $y \in E$ a unique $x \in D$ such that $y = f(x)$. The inverse function is often denoted by $f^{-1}$.

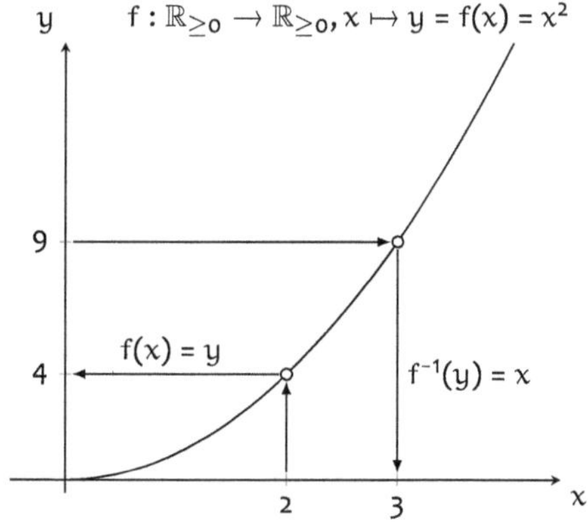


If a function is not injective, it can't be bijective.

$f : \mathbb{R} \to \mathbb{R}, x \mapsto f(x) = x^2$
$(-2)^2 = 2^2 = 4$
(two different x's, same y)
$\Rightarrow$ f not injective
$\Rightarrow$ f not bijective

If a function is not surjective, it can't be bijective.

$f : \mathbb{R}_{\geq 0} \to \mathbb{R}, x \mapsto f(x) = \dfrac{1}{1 + x^2}$
$\Rightarrow f(x) \leq 1$
$\Rightarrow$ There is no x with $f(x) = 2$
$\Rightarrow$ f not surjective
$\Rightarrow$ f not bijective

# How do you get from a linear function to the corresponding line?

The graph of a linear function is a straight line. If you want to draw it, you only need two points.

$f(x) = 2x + 1$
$l : y = 2x + 1$

The easiest way is to choose $x_1 = 0$ as your first x-value.

$y_1 = f(x_1) = f(0) = 2 \cdot 0 + 1 = 1$, so the first point is $P_1(0, 1)$

As your second x-value, you take $x_2 = 1$.

$y_2 = f(x_2) = f(1) = 2 \cdot 1 + 1 = 3$, so the second point is $P_2(1, 3)$

Since the first x-value is zero, your first point $P_1$ lies on the y-axis.

You also plot the second point in the coordinate system.

If you draw a line through the two points, you get the line corresponding to the given equation.

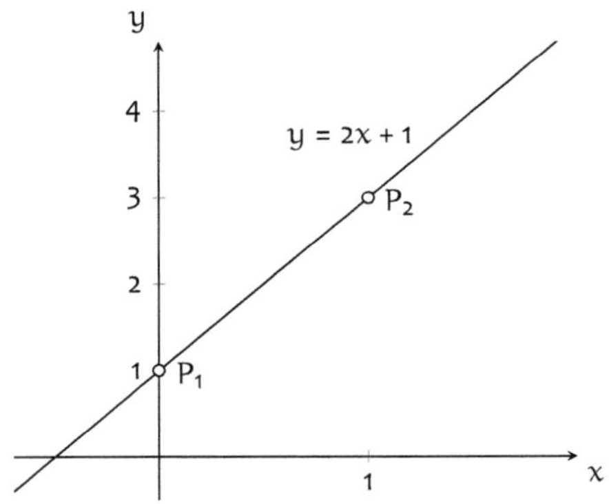


If the coefficient of $x$ is positive, the line runs from bottom left to top right, i.e. it rises. If the coefficient is negative, the line runs from top left to bottom right, it drops.

If the coefficient is zero, the line $y = c$ is parallel to the x-axis: for positive $c$ it lies above the x-axis, for negative $c$ below it.

# What is the slope of a line and where is the slope triangle?

If you take a point on a line and move a bit to the right and then up or down to meet the line again, you obtain a triangle called the slope triangle.

The easiest way is to take $x = 0$ as your first point and move 1 unit to the right.

For the line $y = mx + b$, the slope triangle has the two leg lengths 1 and $m$.

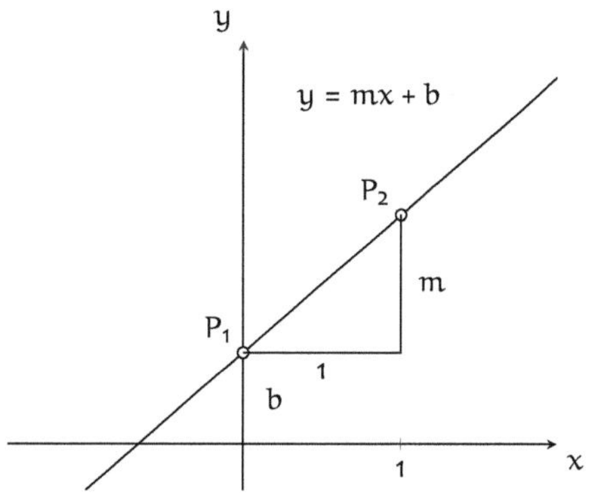


The slope of the line is defined as the ratio of the vertical side to the horizontal side.

$$\frac{m}{1} = m$$

The slope is therefore the tangent of the angle that the line makes with the x-axis.

$$m = \tan(\alpha)$$

You can also use a triangle with other side lengths. If the horizontal side is $\Delta x = x_2 - x_1$ and the vertical side is $\Delta y = y_2 - y_1$, the slope is $m = \Delta y / \Delta x$.

$$m = \frac{\Delta y}{\Delta x} = \frac{y_2 - y_1}{x_2 - x_1}$$

# How else can you get from a linear function to the corresponding line?

You can also substitute another $x$-value than 1 into the equation of the line, for example, if the slope is a fraction, choose an $x$-value that cancels the denominator so that you get a neat, integer-valued $y$.

$$f(x) = -\frac{1}{3}x + 2$$

$$x_1 = 0 \Rightarrow y_1 = 2, \quad P_1(0, 2)$$

$$x_2 = 3 \Rightarrow y_2 = -\frac{1}{3} \cdot 3 + 2 = 1, \quad P_2(3, 1)$$

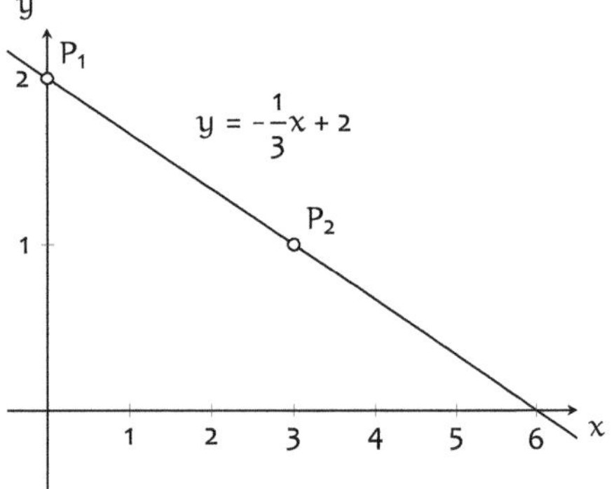


You can also use the intersection with the $x$-axis as the second point. To do this, set $y = 0$ and solve for $x$.

$$f(x) = -\frac{1}{3}x + 2$$

$$x_1 = 0 \Rightarrow y_1 = 2, \quad P_1(0, 2)$$

$$y_2 = -\frac{1}{3}x_2 + 2 = 0$$

$$\frac{1}{3}x_2 = 2 \Rightarrow x_2 = 6, \quad P_2(6, 0)$$

# How do you go from a line in the plane to its equation?

It depends on what information you are given.

If you are given the y-intercept and the slope, you can write down the equation of the line directly in slope-intercept form.

With this, you can describe all lines in the plane except those parallel to the y-axis. The equations of the parallels are $x = a$.

If you are given two points, the easiest way is to substitute both points into the general equation of a line and solve the resulting linear system for m and b.

$y(x) = mx + b$
m is the slope
b is the y-intercept

$x = 0$ is the y-axis
$x = 1$ is the line parallel to the y-axis
through $P(1, 0)$

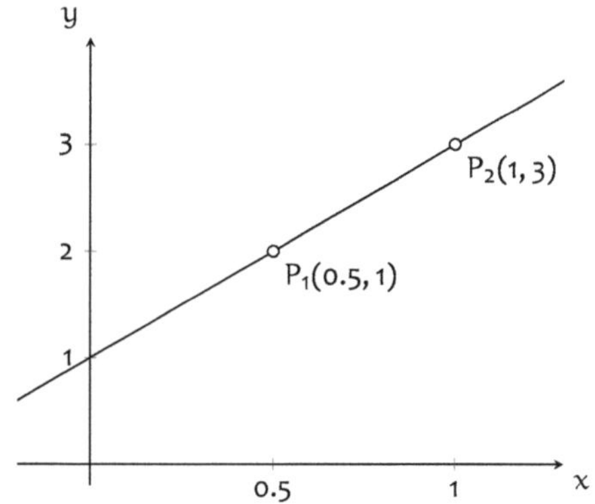


$y = mx + b$
$P(0.5, 2) \Rightarrow 2 = 0.5m + b$
$P(1, 3) \Rightarrow 3 = m + b$

Subtract one equation from the other:
$\Rightarrow 1 = 0.5m \Rightarrow m = 2$
$\Rightarrow 3 = 2 + b \Rightarrow b = 1$

$y = 2x + 1$

# How do you get the intercept form of a line?

If you are given the x-intercept $a$ and the y-intercept $b$, you can write down the intercept form directly.

$$\frac{x}{a} + \frac{y}{b} = 1$$

$a$: x-intercept

$b$: y-intercept

You can write the equations of all lines that are not parallel to either of the two coordinate axes in intercept form.

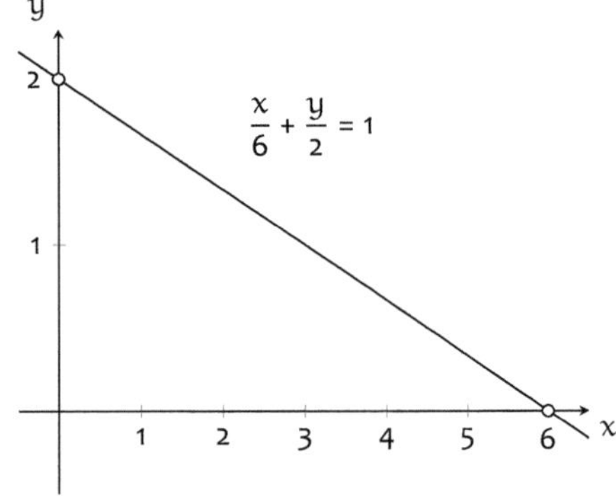


The equations of the lines parallel to the x-axis are $\frac{y}{b} = 1$, you omit the term with $x$, since in a sense the x-intercept $a$ is infinite. Similarly, the equations of the lines parallel to the y-axis are $\frac{x}{a} = 1$.

Parallel to the y-axis:

$$\frac{y}{b} = 1 \Rightarrow y = b$$

Parallel to the x-axis:

$$\frac{x}{a} = 1 \Rightarrow x = a$$

You can of course easily convert the intercept form into the slope-intercept form by solving for $y$.

$$\frac{x}{a} + \frac{y}{b} = 1$$

$$\Rightarrow y = b\left(1 - \frac{x}{a}\right) = -\frac{b}{a}x + b$$

# What is the equation of the standard parabola at the origin and away from the origin?

The simplest quadratic function is $y = x^2$.

| $x$ | -2 | -1 | 0 | 1 | 2 |
|---|---|---|---|---|---|
| $y = x^2$ | 4 | 1 | 0 | 1 | 4 |

The graph is called the standard parabola. The point with the greatest curvature is called the vertex V. For the standard parabola, this is $V = (0, 0)$, the minimum.

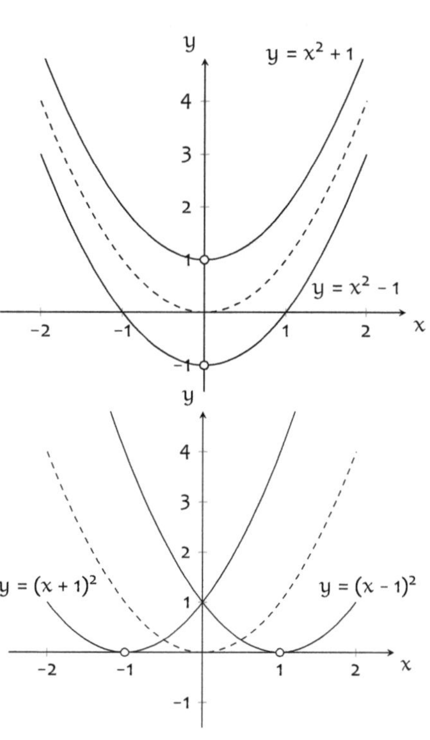


If you add a constant to the formula for the standard parabola, the parabola shifts by this value in the y-direction. If it is positive, the parabola shifts upward; if it is negative, it shifts downward.



If you replace $x$ by $x - a$, the parabola shifts by $a$ to the right: previously, the vertex was at $x = 0$, now it is at $x - a = 0$, that is, $x_V = a$. Note that with $x - \ldots$ you shift to the right. For example, if you replace $x$ by $x + 1 = x - (-1)$, the parabola has its vertex at $x_V = -1$, i.e. you shift to the left.

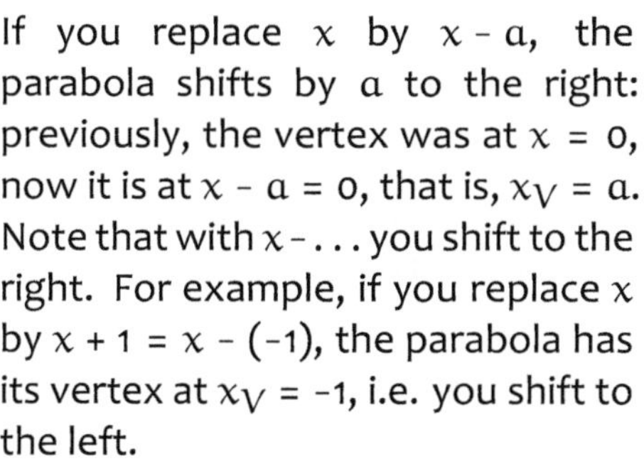

# How do you get from a quadratic function to the corresponding parabola, and vice versa?

The standard parabola with vertex at $V = (a, b)$ is shifted by $a$ to the right and by $b$ upwards, so its equation is $y = (x - a)^2 + b$.

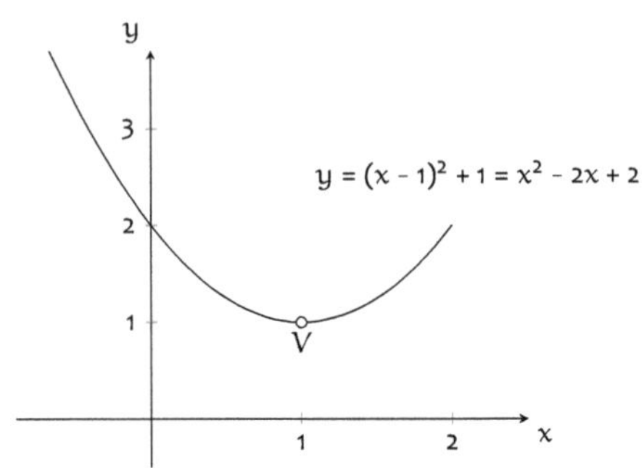



If you have an equation of the form $y = x^2 + px + q$, you can complete the square to bring it into the form $y = (x - a)^2 + b$. Then you can directly read off the coordinates of the vertex $V(a, b)$.

$$y = x^2 - 2x + 2 = x^2 - 2x + 1 - 1 + 2$$
$$\Rightarrow y = (x - 1)^2 + 1$$
$$\Rightarrow \text{vertex } V(1, 1)$$

In addition to the shift, a parabola can also be stretched or compressed in the $y$-direction. The equation of the stretched or compressed standard parabola is $y = kx^2$. For $k > 1$ it is stretched, for $0 < k < 1$ it is compressed, for $k < 0$ it is reflected about the $y$-axis and, depending on $|k|$, compressed or stretched.

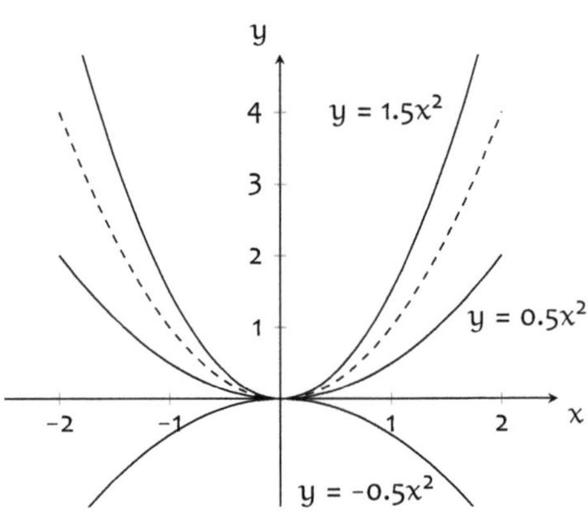



To graph a quadratic function $y = Ax^2 + Bx + C$, you bring it into the form $y = k(x - x_V)^2 + y_V$. To do this, first factor out $A$ and then proceed as above. You can then read off the vertex $V = (x_V, y_V)$ and the stretch factor $k = A$.

$$y = 2x^2 - 8x - 6$$
$$= 2(x^2 - 4x - 3)$$
$$= 2(x^2 - 4x + 4 - 4 + 3)$$
$$= 2(x - 2)^2 - 2$$
$$\Rightarrow \text{vertex } V(2, 2), \text{ stretch factor } 2$$

# How do you determine whether a function is axially symmetric or point-symmetric?

For a function symmetric with respect to the y-axis, the y-value is identical when you replace x by –x.

$f(-x) = f(x)$
$\Rightarrow$ axially symmetric w.r.t. the y-axis
$f(x) = x^2 - x^4$:
$f(-x) = (-x)^2 - (-x)^4 = x^2 - x^4 = f(x)$

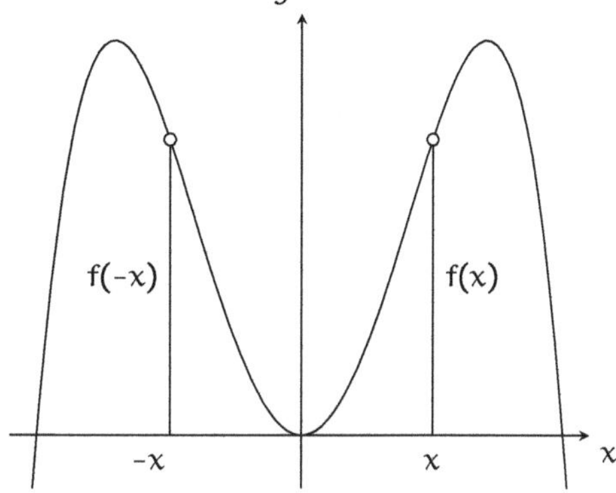


For a function that is point-symmetric with respect to the origin, the y-value changes its sign and keeps the same absolute value when you replace x by –x.

$f(-x) = -f(x)$
$\Rightarrow$ point-symmetric w.r.t. the origin
$f(x) = x - x^3$:
$f(-x) = (-x) - (-x)^3 = -x + x^3 = -f(x)$

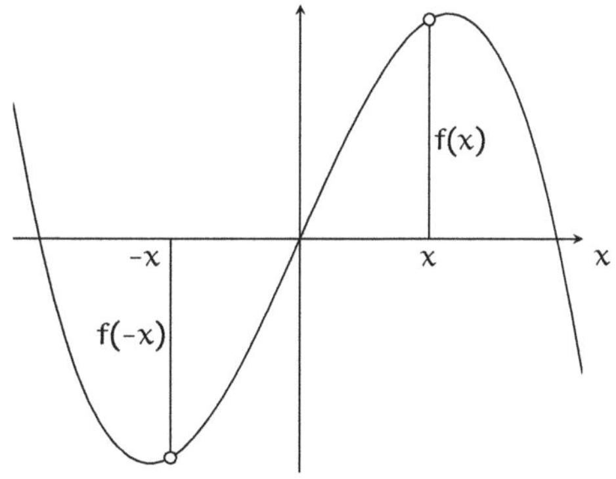


Similarly, you can determine whether a function is axially symmetric w.r.t $x = a$ or point-symmetric w.r.t $P(a, b)$.

$f(2a - x) = f(x)$
$\Rightarrow$ axially symmetric w.r.t. $x = a$
$f(2a - x) = 2b - f(x)$
$\Rightarrow$ point-symmetric with respect to $P(a, b)$

# How do you shift functions?

You shift arbitrary functions in exactly the same way as parabolas.

If you add a constant to the function, the graph shifts by that value in y-direction: upward if the value is positive, downward if it is negative.

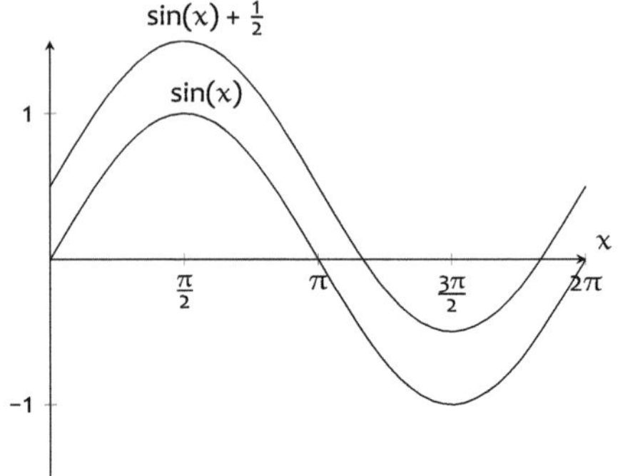


If you replace $x$ in the argument of the function by $x - a$, the graph shifts by $a$ units to the right (pay attention to the sign), since the original position $x = 0$ is now the new position $x - a = 0$, that is, $x = a$.

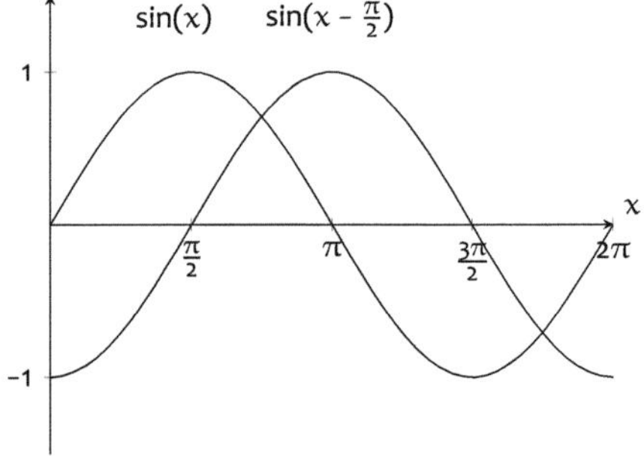

# How do you stretch functions in the y-direction?

To stretch a function in the y-direction, you multiply the values by a factor greater than 1.

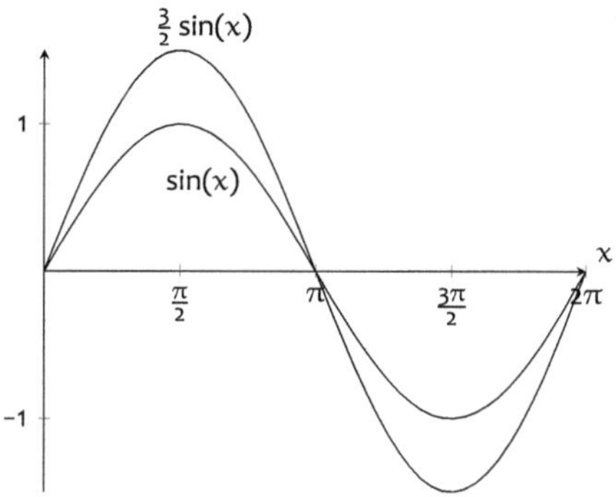


If the factor is less than 1, the function is compressed in the y-direction.

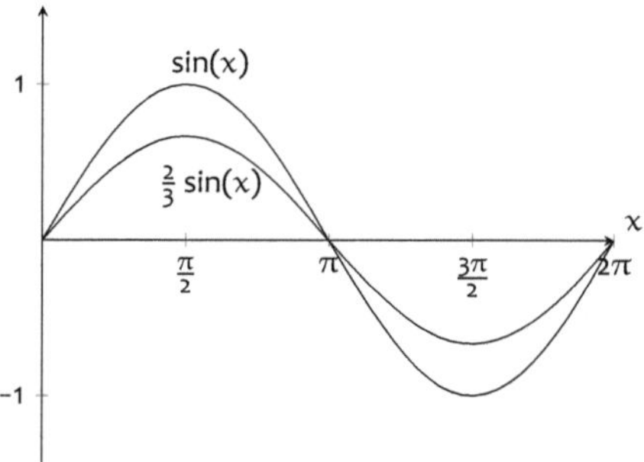


If the factor is negative, the function is reflected across the x-axis and stretched or compressed in the y-direction by the absolute value of the factor.

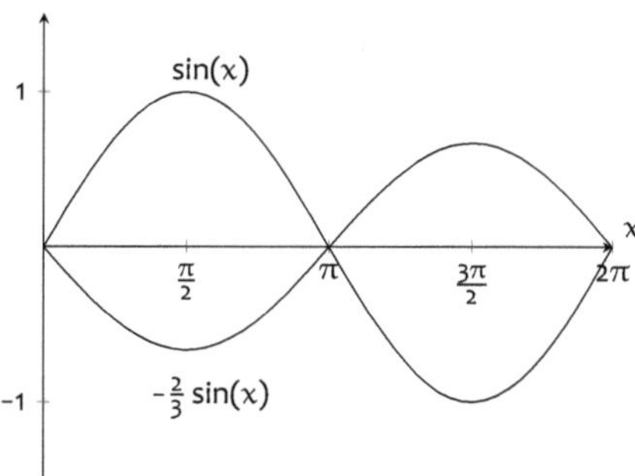

# How do you stretch functions in the $x$-direction?

To stretch a function in the $x$-direction by a factor $a > 1$, you replace $x$ by $x/a$, i.e. you have to divide $x$ by the factor.

On the right side $a = 2$ which gives $\frac{x}{2}$ as argument of the stretched function. This is similar to the shift, where you have to subtract the amount by which you shift to the right from $x$.

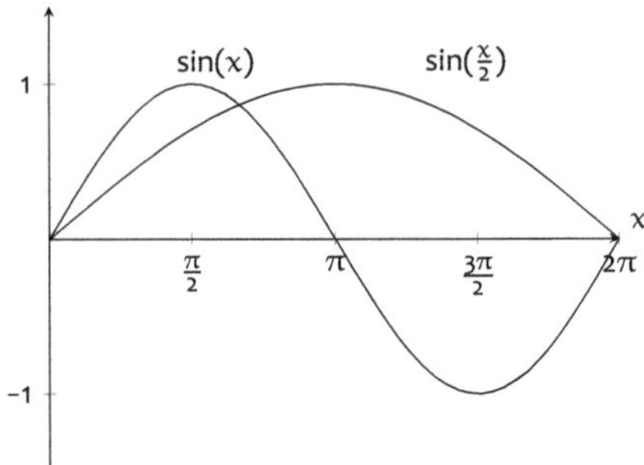


To compress a function in the $x$-direction, you replace $x$ with $x/a$ with a factor $a < 1$.

On the right side $a = \frac{1}{2}$ which gives $2x$ as argument of the compressed function.

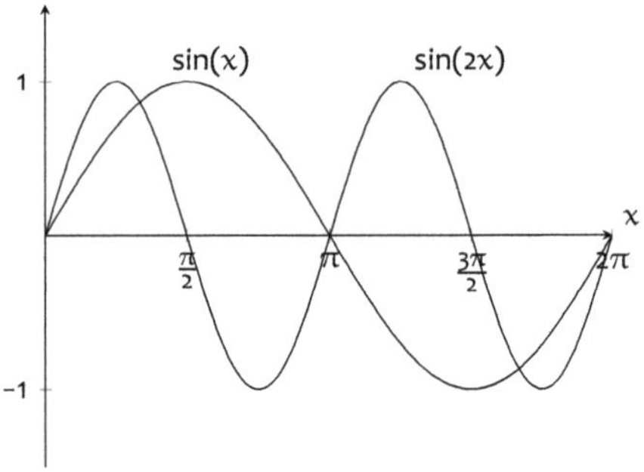


For graphs of power functions, stretches/compressions in the $x$- and $y$-directions can be converted into each other. For a parabola, a stretch by $a$ in the $y$-direction corresponds to a compression by $1/\sqrt{a}$ in the $x$-direction.

$y = x^2$

$y = 2x^2$: stretching by a factor of 2 in the $y$-direction

$y = (x/(1/\sqrt{2}))^2 = 2x^2$: compression by a factor of $1/\sqrt{2}$ in the $x$-direction

# How do you reflect functions over the line $y = x$?

The reflection of the $x$- and $y$-axes over the line $y = x$ means that the $x$- and $y$-axes are interchanged, and thus also the $x$- and $y$-coordinates of the points.

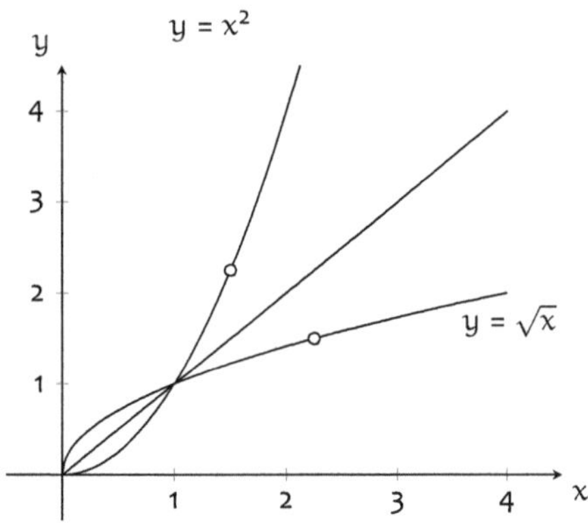


Therefore, to obtain the function reflected over the line $y = x$, you must swap the variables $x$ and $y$ in the equation $y = f(x)$ and then solve for $y$ again.

$$f(x) = y = x^2 \quad | \text{ swap } x \text{ and } y.$$
$$x = y^2 \quad | \quad \sqrt{\ } \text{ solve for } y.$$
$$\sqrt{x} = y \quad | \text{ write } y \text{ on the left.}$$
$$y = \sqrt{x} = f^{-1}(x)\text{: inverse function}$$

The function reflected over the line $y = x$ is called the inverse function, and it is often denoted by $f^{-1}$.

$f^{-1}$ Inverse function:
$$f^{-1}(f(x)) = f(f^{-1}) = x$$
The notation can be misleading: it is not the function to the power of –1, i.e. $1/f$.

In order to form the inverse function, the original function must be monotonic, which can be achieved if necessary by restricting the domain.

$x^2$ can be inverted for $x > 0$, not for $x \in \mathbb{R}$.
$\sin(x)$ can be inverted for $x \in [-\frac{\pi}{2}, \frac{\pi}{2}]$, not for $x \in \mathbb{R}$.

Some inverse functions are needed often and even have their own names.

| $f(x)$ | $f^{-1}(x)$ |
| --- | --- |
| $x^2$ | $\sqrt{x}$ |
| $x^n$ | $x^{1/n} = \sqrt[n]{x}$ |
| $1/x$ | $1/x$ |
| $a^x$ | $\log_a(x)$ |
| $e^x$ | $\ln(x)$ |
| $\sin(x)$ | $\arcsin(x)$ |

# When are two lines perpendicular to each other?

If you reflect a line with slope $m$ over the bisector of the first quadrant, i.e. $y = x$, then the slope of the reflected line is the reciprocal of the original slope: $m' = 1/m$.

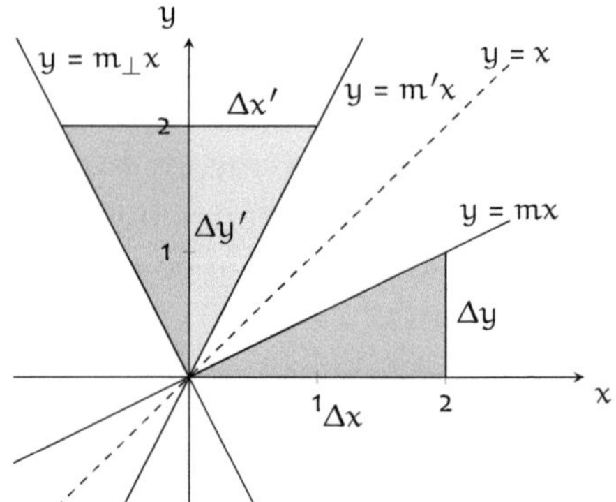


You can see that if you look at the slope triangles: the adjacent and opposite sides, i.e. $\Delta x$ and $\Delta y$, are interchanged.

$$m' = \frac{\Delta y'}{\Delta x'} = \frac{\Delta x}{\Delta y} = \frac{1}{\frac{\Delta y}{\Delta x}} = \frac{1}{m}$$

You can also interchange $x$ and $y$ in the equation of the line, which is how you form the inverse function. If you then solve for $y$ again, you can read off the new slope.

$$y = mx$$
$$x = my$$
$$my = x$$
$$y = \frac{1}{m}x = m'x \quad \text{with} \quad m' = \frac{1}{m}$$

If you then reflect the line over the $y$-axis, you get a line that is perpendicular to the original one.

This holds because two successive reflections over axes that intersect at an angle of $\alpha = 45°$ result in a rotation by $2\alpha = 90°$.

If you reflect over the $y$-axis, $x$ becomes $-x$, so $\Delta x$ also becomes $-\Delta x$. This gives you the slope of the perpendicular line.

$$m_\perp = \frac{\Delta y_\perp}{\Delta x_\perp} = \frac{\Delta y'}{-\Delta x'} = -m' = -\frac{1}{m}$$

E.g.

$$m = \frac{1}{2}, \ m' = \frac{1}{m} = 2 \text{ and } m_\perp = -2.$$

The easiest way to remember this is that the product of the slopes of two perpendicular lines is always $-1$.

$$mm_\perp = -1$$

# What do power functions look like?

The power functions $x^n$ for even $n$ are symmetric with respect to the $y$-axis; they come from the upper left, have a minimum at $x = 0$, and then go up to the right. The larger $n$ is, the steeper the curve. All functions pass through the points $(-1, 1)$ and $(1, 1)$.

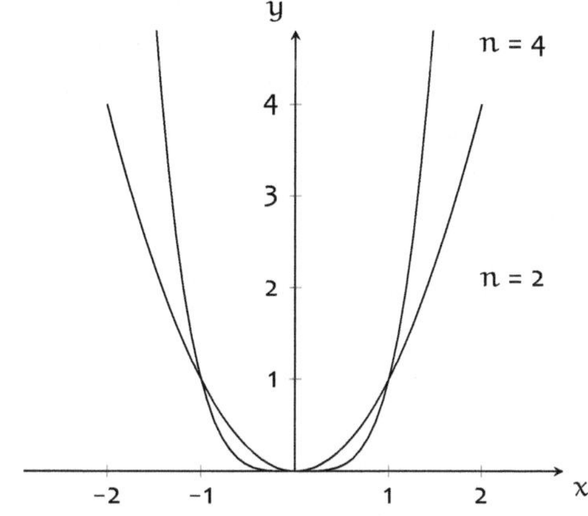


The power functions $x^n$ for odd $n$ are point-symmetric with respect to the origin; they come from the bottom left, have a saddle point at $x = 0$, and then go up to the top right. The larger $n$ is, the steeper the curve. All functions pass through the points $(-1, -1)$ and $(1, 1)$.

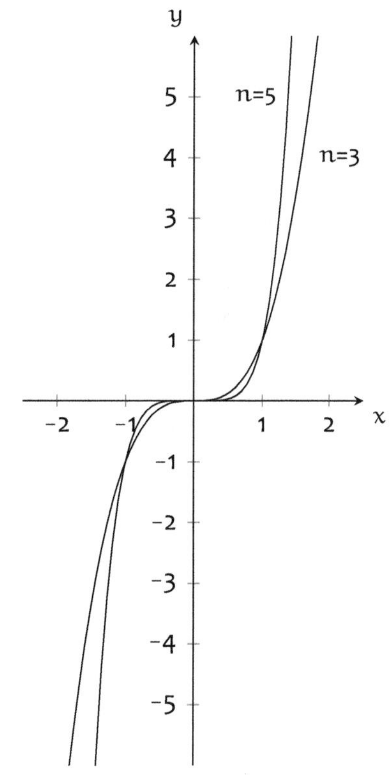

# What is a polynomial?

A polynomial also known as polynomial function is a sum of powers of an unknown with non-negative integer exponents, each multiplied by a constant (called a coefficient).

All kinds of coefficients, be they zero or negative or decimal numbers, are allowed. You can have any number of different terms.

It's important that a polynomial must not contain negative or fractional exponents.

The exponent of the highest power is called the degree of the polynomial.

A polynomial can also be written as a product of factors.

If you have complex numbers available, you can write any polynomial as a product of linear factors, from which you can directly read off the zeros.

If you only use real numbers, there are polynomials that can't be split into linear factors. However, every polynomial over $\mathbb{R}$ can be written as a product of at most quadratic factors.

$$p(x) = a_0 + a_1 x + a_2 x^2 + \ldots + a_n x^n$$

$$p(x) = \sum_{i=0}^{n} a_i x^i$$

examples of polynomials:
$$p_1(x) = x^3 - 6x^2 + 11x - 6$$
$$p_2(x) = x^5 + x^{10}$$
$$p_3(x) = 0.234x - 17.12x^6 + 7x^7$$

$\frac{1}{x} + x^2$ is not a polynomial, since
$\frac{1}{x} = x^{-1}$ and $-1$ is negative.
$\sqrt{x} + 5x$ is not a polynomial, since
$\sqrt{x} = x^{1/2}$ and $1/2$ is a fraction.

$5 - 3x$ has degree 1.
$x^5 + x^{10}$ has degree 10.

$$x^3 - 6x^2 + 11x - 6 = (x - 1)(x - 2)(x - 3)$$

$$x^2 + 3x + 2 = (x + 1)(x + 2)$$
$$x^2 - 1 = (x + 1)(x - 1)$$
$$x^2 + 1 = (x + i)(x - i)$$
$$x^4 + 4 =$$
$$= (x + 1 + i)(x + 1 - i)(x - 1 + i)(x - 1 - i)$$

$x^2 + 1 > 0$ for $x \in \mathbb{R}$
can't be factored over $\mathbb{R}$.

$$x^4 + 4 =$$
$$= (x^2 + 2x + 2)(x^2 - 2x + 2)$$
can't be further factored over $\mathbb{R}$.

# How do you find the zeros of polynomials?

You often need to find the zeros of polynomials, for example to sketch a polynomial.

The zero of a function is an $x$-value for which the $y$-value, $y = f(x)$, is zero. That is, where the graph of the function crosses the $x$-axis. On the right, the zeros are $x = 0$ and $x = 1$.

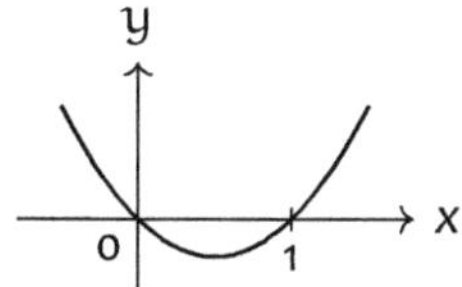


How do you find zeros?

If your polynomial is already factored: leave it that way, do not expand. You can then simply read off the zeros.

zeros of $f(x) = x(x - 1)$:
$x(x - 1) = 0$
$\Rightarrow x = 0$ or $x - 1 = 0$
$\Rightarrow x_1 = 0, x_2 = 1$
are the two zeros.

If your polynomial is of degree one, i.e. a linear function: just solve the linear equation.

zero of $f(x) = 3x + 5$:
$3x + 5 = 0$
$\Rightarrow 3x = -5$
$\Rightarrow x = -\frac{5}{3}$

If your polynomial is of second degree, i.e. a quadratic polynomial: solve the quadratic equation.

zeros of $f(x) = x^2 - 5x + 6$:
$x^2 - 5x + 6 = 0$
$\Rightarrow x_{1,2} = \frac{5 \pm \sqrt{25-24}}{2}$
$\Rightarrow x_{1,2} = \frac{5 \pm 1}{2}$
$\Rightarrow x_1 = 3, x_2 = 2$

If your polynomial is of higher degree, see the section on solving higher degree equations for exact methods. Another possibility are numerical approximation methods, some of which are also implemented in calculators.

# What are multiple zeros?

If you multiply two functions, each of which has a zero, but at different points, you get a function that has a zero at both points, since 0 times any number is still 0.

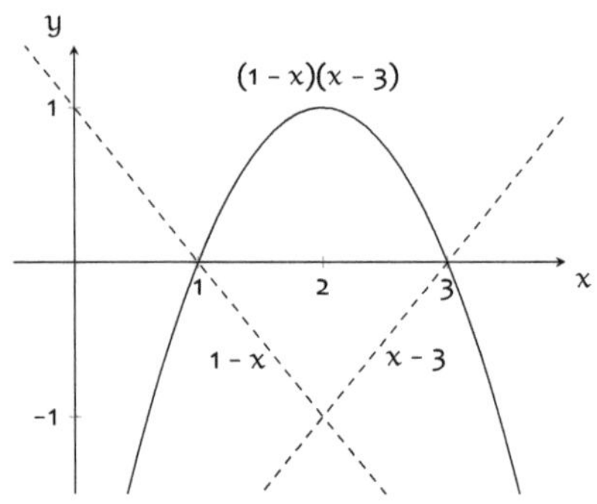



a zero at $x = 1$ and $x = 3$

If you multiply two functions that have a zero at the same point, you get a function that has, as we say, a double zero at that point. You can imagine that this happened because the two zeros have merged together.

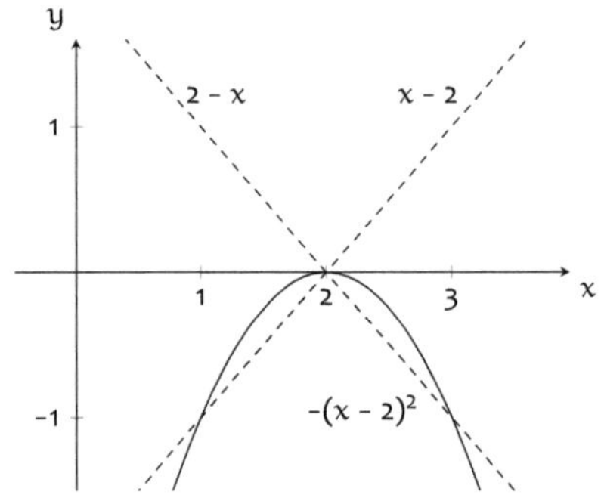



a double zero at $x = 2$

Similarly, there are also triple, quadruple, and higher order zeros.

$$(x + 1)^3(x - 2)^2$$
has a triple zero at $x = -1$

For a zero of order $n$, the function locally looks like $x^n$ (or $-x^n$), i.e. for even $n$ it does not change signs, for odd $n$ it does change signs, and the larger $n$ is, the flatter it is.

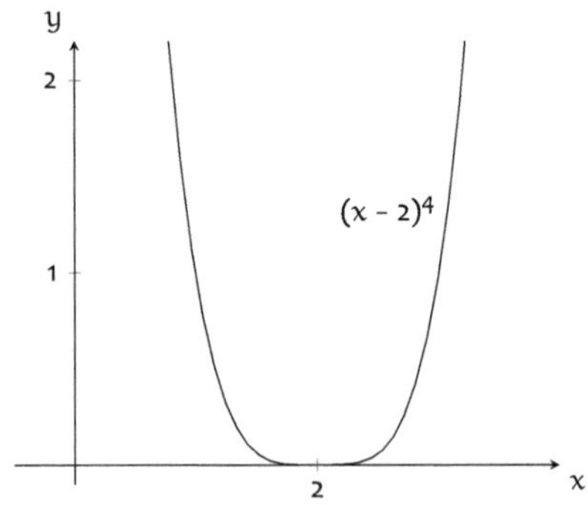



a quadruple zero at $x = 2$

# What are rational functions?

Rational functions are the quotient of two polynomials. They are called rational because a quotient is also known as a ratio.

$$y = f(x) = \frac{x^2}{x - 1}$$

For rational functions, you should note the following:

- The function is zero when the numerator is zero. Thus we get exactly the same problem as for polynomials.

$$f(x) = \frac{x^2}{x - 1} = 0 \Rightarrow x^2 = 0 \Rightarrow x = 0$$

- The function has holes in the domain whenever the denominator is zero, since division by zero is not allowed.

$$x - 1 = 0 \Rightarrow x = 1$$
$$f(x) \text{ is defined for } x \in \mathbb{R} \setminus \{1\}.$$

- If the numerator has no zero where the denominator has a zero, the function has a vertical asymptote at that point, and the function value to the right and left of it approaches $+\infty$ or $-\infty$.

$$\lim_{\substack{x \to 1 \\ x<1}} \frac{x^2}{x - 1} = -\infty$$
$$\lim_{\substack{x \to 1 \\ x>1}} \frac{x^2}{x - 1} = +\infty$$

- When differentiating, you use the quotient rule.

$$f'(x) = \frac{2x \cdot (x - 1) - 1 \cdot x^2}{(x - 1)^2} = \frac{x^2 - 2x}{(x - 1)^2}$$

- To find the behavior for $x \to \pm\infty$, you use polynomial division and calculate the limit.

$$f(x) = \frac{x^2}{x - 1} = x + 1 + \frac{1}{x - 1}$$

The graph of f approaches the line $x + 1$ for $x \to \pm\infty$.

# How does polynomial division work?

If a polynomial $p(x)$ has a zero at $x = a$, then it can be written as $p(x) = (x - a)q(x)$, where $q(x)$ is a polynomial of degree one fewer.

$x^2 - 3x + 2$ (of degree 2) has a zero at $x = 1$.

Therefore:
$x^2 - 3x + 2 = (x - 1)\, q(x)$
with $q(x) = x - 2$    (of degree 1)

To find the polynomial $q(x)$, you have to divide $p(x)$ by $x - a$, which is called polynomial division.

$$p(x) = (x - a)\, q(x) \Rightarrow q(x) = \frac{p(x)}{x - a}$$

If, for example, you want to find the zeros of $x^3 - 6x^2 + 11x - 6$, you can find a solution $x = 1$ by guessing (or by trying the divisors of the constant term $-6$).

$x^3 - 6x^2 + 11x - 6 = 0$
has a solution $x = 1$.
Therefore:
polynomial division by $x - 1$:

With polynomial division, you proceed similarly to long division. In this example, you look at how many times $x - 1$ fits into $x^3$, considering only the highest power $x$ at first; the answer is $x^2$ times. Then you multiply this $x^2$ by $x - 1$, obtaining $x^3 - x^2$, which you subtract. Then you continue in the same way.

$$
\begin{array}{l}
\left(\ x^3 - 6x^2 + 11x - 6\right) \div (x - 1) = x^2 - 5x + 6 \\
\underline{-\,x^3\ \ \ + x^2} \\
\qquad -5x^2 + 11x \\
\qquad \underline{5x^2 - 5x} \\
\qquad\qquad 6x - 6 \\
\qquad\qquad \underline{-6x + 6} \\
\qquad\qquad\qquad 0
\end{array}
$$

By the end, it should balance out exactly, leaving a zero in the last row. As a result of the division, you obtain a quadratic polynomial.

$(x^3 - 6x^2 + 11x - 6) \div (x - 1) = x^2 - 5x + 6$
$\Rightarrow x^3 - 6x^2 + 11x - 6 = (x - 1)(x^2 - 5x + 6)$

You can find its zeros as usual.

$x^2 - 5x + 6 = 0 \Rightarrow x = 2, x = 3$

Thus, you obtain all zeros of the cubic polynomial.

$x^3 - 6x^2 + 11x - 6 \Rightarrow x = 1, x = 2, x = 3$

# How does polynomial division with remainder work?

For a rational function where the degree of the numerator is at least as large as the degree of the denominator, you can use polynomial division.

$$\begin{array}{l}(\phantom{-}x^2) \div (x-1) = x + 1 + \dfrac{1}{x-1}\\[2pt] \underline{-x^2 + x}\\[2pt] \phantom{--}x\\[2pt] \phantom{--}\underline{-x+1}\\[2pt] \phantom{----}1\end{array}$$

In general, it will not divide evenly. You then write the remainder as a fraction. This representation shows you the behavior for $x \to \pm\infty$.

The graph of the function

$$f(x) = \frac{x^2}{x-1} = x + 1 + \frac{1}{x-1}$$

approaches the line $x+1$ for $x \to \pm\infty$.

Lines that the graph of the function approaches arbitrarily closely are called asymptotes.
$x = 1$ is a vertical asymptote,
$y = x + 1$ is an oblique asymptote.

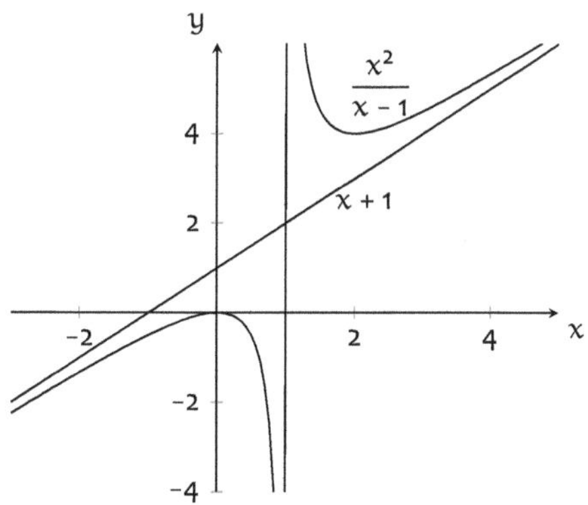


Functions other than rational functions can also have asymptotes.

$f(x) = \sqrt{1 + x^2}$ has the asymptote $y = x$ for $x \to \pm\infty$.

$f(x) = 1 + e^x$ has the asymptote $y = 1$ for $x \to -\infty$.

# What do exponential functions look like?

The exponential function $a^x$ for $a > 1$ starts for negative $x$ at small $y$-values , passes through the point $(0, 1)$ since $a^0 = 1$ for all $a \neq 0$ , and goes to $\infty$ for positive $x$. The larger $a$ is, the steeper the curve.

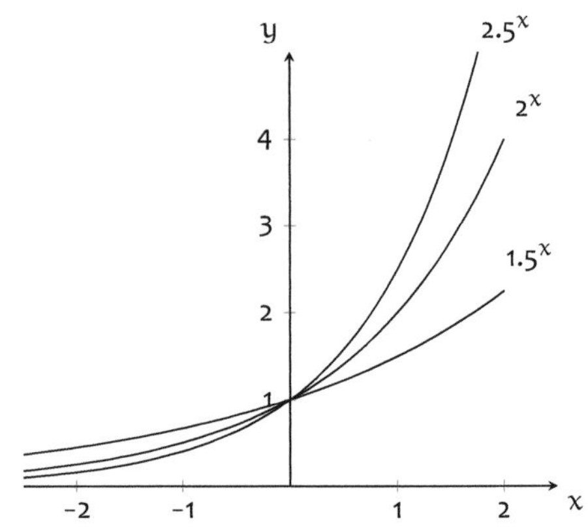


The exponential function $a^x$ for $a < 1$ starts for negative $x$ at large $y$-values , passes through the point $(0, 1)$ since $a^0 = 1$, and goes to zero for positive $x$. The positive $x$-axis is the asymptote for $x \to \infty$.

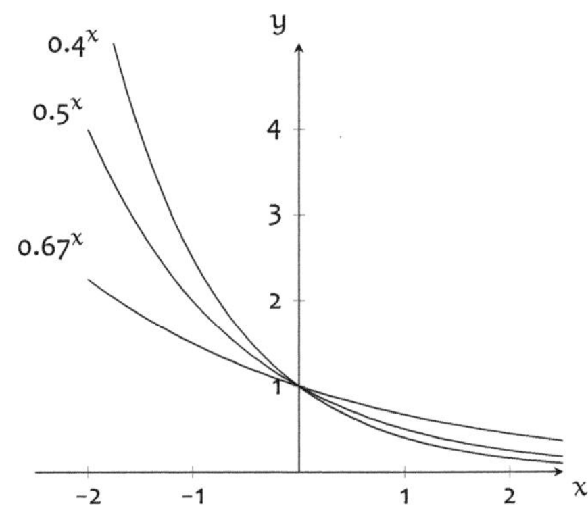


Because $a^{-x} = \left(\frac{1}{a}\right)^x$, the curve $2^x$ is reflected over the $y$-axis to become the curve $0.5^x$.

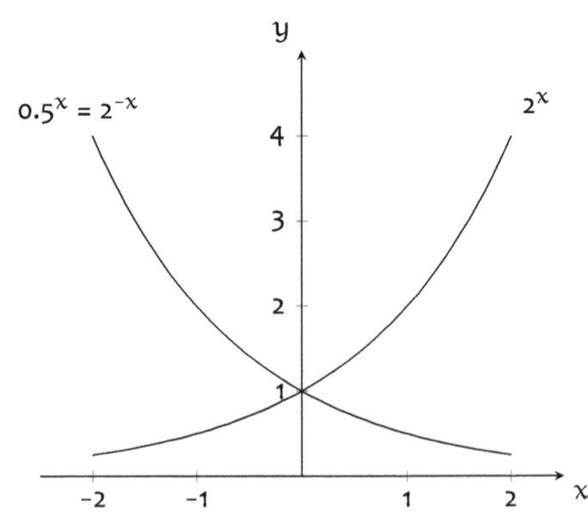

# What do logarithmic functions look like?

The logarithmic functions $\log_a(x)$ are the inverse functions of $a^x$: They are only defined for $x > 0$. They start with negative function values for small $x$, pass through the point $(1, 0)$, since $\log_a(1) = 0$ for all $a > 0$, and tend to $\infty$ for positive $x$, but very slowly, slower than any root function.

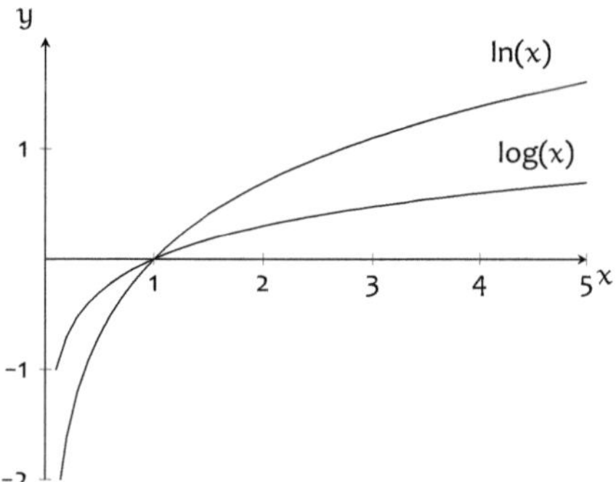

For $x \rightarrow 0$ the logarithm tends to $-\infty$, the graph of the logarithm functions has the vertical asymptote $x = 0$.

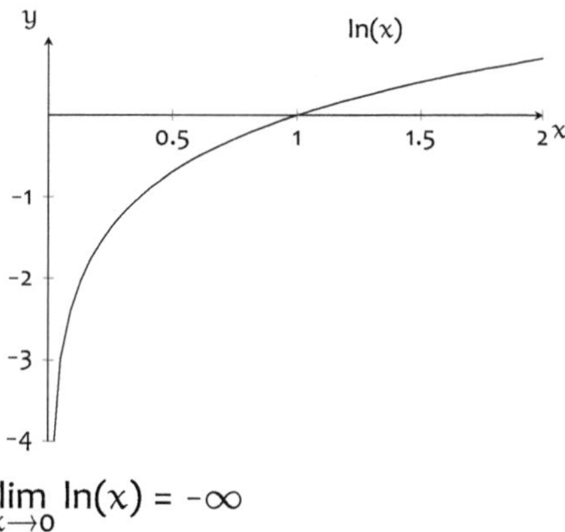

$$\lim_{x \rightarrow 0} \ln(x) = -\infty$$

The logarithm is only defined for positive values. This restricts the domain of the logarithm functions.

$f(x) = \ln(x),\ x > 0$

$g(x) = \ln(10 - 2x),\ x < 5$

# Overview: Injective, surjective, and bijective

| | Description and Example | Counterexample |
|---|---|---|
| Function | Every x-value has exactly one y-value. | If there is more than one or no y-value for an x-value, it is not a function. |
| Injective | Every y-value has at most one x-value. | If there is more than one x-value for a y-value, the function is not injective. |
| Surjective | Every y-value has at least one x-value. | If there is no x-value for a y-value, the function is not surjective. |
| Bijective | Every y-value has exactly one x-value.<br><br>You can reverse the arrows and obtain the inverse function. | If there is more than one or no x-value for a y-value, the function is not bijective.<br><br>The function does not have an inverse function. |

# Sequences and Limits

We now turn to analysis, a branch of higher mathematics that was founded in the 17th century by Leibniz and Newton.

Analysis can be understood as the taming of the infinitely small and the infinitely large.

Taming means that you can calculate with infinitely small and infinitely large quantities and still obtain finite quantities.

The finite quantity could, for example, be the slope of a curve at a given point. For this you would need to calculate the slope of an infinitesimal triangle. This would give you $\frac{0}{0}$ which you can't calculate directly but only via a limiting process.

Another example would be the area under a curve, which can be imagined as being composed of infinitely many infinitely thin rectangles.

To define these concepts precisely, infinite sequences and their limits have proven useful, which is what we will deal with first.



A. Gründers, *Math Made Clear: From the Basics to Calculus*, https://doi.org/10.1007/978-3-662-73221-2_8

# What are indices used for?

If you have different variables that denote similar things, you can also use a numbered variable, also known as indexed variable.

Instead of $a, b, c, \ldots$ simply use $a_1, a_2, a_3, \ldots$

The number of the variable is written at the lower right and is called the index.

$a_3 \leftarrow$ index

If you want to consider all these variables at once, you can use a variable (often $i$, $j$, or $n$) as the index.

$a_i$

You have to specify which values the index variable can take.

With $a_i$, $i = 1, 2, 3$, $a_1, a_2, a_3$ is meant.

This can be used to denote $n$ variables at the same time.

With $a_i$, $i = 1, 2, \ldots, n$ $a_1, a_2, \ldots, a_n$ is meant.

# How do you calculate with summation signs?

If you want to add all variables from a certain index range, you can write this with a summation sign, which is the Greek capital letter sigma.

$$a_1 + a_2 + a_3 + a_4 = \sum_{i=1}^{4} a_i$$

You write the first index value below the summation sign, i.e. as subscript. Similarly, you write the last index above the summation sign, i.e. as superscript.

$$\sum_{i=1}^{4} \quad \substack{4 \leftarrow \text{upper index value} \\ \\ i=1 \leftarrow \text{lower index value}}$$

The first and last values are also called the limits of summation.

It doesn't matter how you name the index variable. Only consistency within a sum is important.

$$\sum_{i=3}^{7} a_i = \sum_{n=3}^{7} a_n = a_3 + \ldots + a_7$$

If you want to shift the summation limits (i.e. summing over $j = i - 3 = 0, \ldots, 4$ instead of $i = 3, \ldots, 7$), you must also adjust the index accordingly.

$$\sum_{i=3}^{7} a_i = \sum_{j=0}^{4} a_{j+3}$$

You can then rename the index variable again.

$$\sum_{i=3}^{7} a_i = \sum_{j=0}^{4} a_{j+3} = \sum_{i=0}^{4} a_{i+3}$$

Of course, you can also sum over a specific expression that depends on $i$ instead of $a_i$.

$$\sum_{i=1}^{3} i^2 = 1 + 4 + 9 = 14$$

You can compact formulas with the summation sign.

$1 + 3 + 5 + 7 = 16.$ i.e.:

$$\sum_{i=1}^{4} (2i - 1) = 4^2 = 16, \text{ in general:}$$

$$\sum_{i=1}^{n} (2i - 1) = n^2$$

# What is the binomial formula for higher powers?

You can expand $(a + b)^n$ for exponents $n > 2$ in a similar way as for $n = 2$.

This results in a sum of $n + 1$ terms $a^k b^{n-k}$, $k = 0, 1, \ldots, n$, where the sum of the exponents, in each term, is equal to $k + (n - k) = n$.

To determine the coefficient of the term $a^k b^{n-k}$, consider how often $a$ appears $k$ times and $b$ appears $(n - k)$ times, when expanding the $n$ parentheses.

The easiest way to think about this is: you can arrange $n$ arbitrary symbols in $n! = 1 \cdot 2 \cdot 3 \cdot \ldots \cdot n$ ways. $n!$ is called $n$ factorial.

If $k$ of the $n$ symbols are $a$ and $n - k$ of the symbols are $b$, and these are indistinguishable, you must divide by the respective number of identical arrangements. i.e. $k!$ and $(n-k)!$.

This expression is called the binomial coefficient.

Thus, you obtain the general formula for $(a + b)^n$, which is also called the binomial theorem.

$$(a + b)^2 = a^2 + 2ab + b^2$$
$$(a + b)^3 = a^3 + 3a^2b + 3ab^2 + b^3$$
$$(a + b)^4 = a^4 + 4a^3b + 6a^2b^2 + 4ab^3 + b^4$$

$$(a + b)^n = \underbrace{a^n + \ldots + b^n}_{n+1 \text{ terms}}$$

of the form $a^k b^{n-k}$, $k=0,\ldots,n$

$$(a + b)^n = (a + b)(a + b)\cdots(a + b)$$

Choose $a$ from $k$ of the $n$ parentheses and $b$ from $n - k$ parentheses. For $n = 4$ and $k = 2$ there are 6 possibilities: $aabb + abab + abba + baab + baba + bbaa = 6a^2b^2$

For the first symbol you have $n$ possibilities, for the second $n - 1$, for the third $n - 2$, and for the last only one possibility. Thus, in total you have $n(n - 1)(n - 2) \cdot \ldots \cdot 1 = n!$ possibilities.

Number of possibilities for $a$ $k$ times and $b$ $(n - k)$ times:

$$\frac{n!}{k!\,(n - k)!} = \frac{n(n - 1)\cdots(n - k + 1)}{1 \cdot 2 \cdots k}$$

$$\binom{n}{k} = \frac{n!}{k!\,(n - k)!}$$

$$(a + b)^n = \sum_{k=0}^{n} \binom{n}{k} a^k b^{n-k}$$

# What is a sequence?

A sequence of numbers is a set of numbers that is arranged in a fixed order.

$1, 2, 3, \ldots$
$10, 8, 6, \ldots$

A sequence is written in parentheses to make it clear that the order matters. This is in contrast to curly brackets for sets, where the order of elements does not matter.

$(a_n) = (1, 2, 3, \ldots)$

Usually, only (countably) infinite sequences are called sequences.

A (real) sequence is thus a mapping from the natural numbers to the set of (real) numbers. The fact that the argument is written as an index is just a notational convention.

$$a : \mathbb{N} \to \mathbb{R}$$
$$n \mapsto a(n) = a_n$$

You can specify the sequence either explicitly (depending only on the index) or recursively (depending on previous terms of the sequence).

$$a_n = n^2 - 2n$$

$$b_n = 2b_{n-1} + n, \ b_1 = 1$$

If the sequence is given explicitly, you can calculate any term of the sequence (i.e. $a_n$ for any $n$) directly.

for $n = 3$ : $a_3 = 3^2 - 3 \cdot 2 = 3$

If the sequence is given recursively, you must calculate the terms of the sequence one after another in order.

$b_1 = 1$ (given)
$b_2 = 2b_1 + 2 = 2 \cdot 1 + 2 = 4$
$b_3 = 2b_2 + 3 = 2 \cdot 4 + 3 = 11$

# How can you prove a statement for all natural numbers?

If you observe a pattern for e.g. $1, 2, 3, 4$ and you wonder whether this holds in general (i.e. for all natural numbers) you can proceed as follows.

Show that, if it holds for an arbitrary but fixed number $n$, then it also holds for the next number $n + 1$. In order to do so, you may assume that the statement is correct for a given $n$.

If you also know that the statement holds for $n = 1$, then you have shown that it holds for all $n \in \mathbb{N}$.

This is called the principle of mathematical induction: you have to show 1) that $A(1)$ holds and 2) that from the statement $A(n)$ the statement $A(n + 1)$ follows.

The statement above can also be easily proven directly.

However, it's often the case that a statement can be proven much more easily by induction than directly, or that it can only be proven in this way.

$$1 + 2 = 3 = \tfrac{1}{2} \cdot 2 \cdot 3$$
$$1 + 2 + 3 = 6 = \tfrac{1}{2} \cdot 3 \cdot 4$$
$$1 + 2 + 3 + 4 = 10 = \tfrac{1}{2} \cdot 4 \cdot 5$$

Is $1 + 2 + \cdots + n = \tfrac{1}{2}n(n + 1)$ true?

Assumption:
$$1 + 2 + \cdots + n = \tfrac{1}{2}n(n + 1)$$
Add $n + 1$ to both sides:
$$1 + 2 + \cdots + n + n + 1 = \tfrac{1}{2}n(n + 1) + n + 1$$

The right side is $= \tfrac{1}{2}(n(n + 1) + 2(n + 1)) = \tfrac{1}{2}(n + 1)(n + 2)$.
This is the statement for $n + 1$.

$1 = \tfrac{1}{2} \cdot 1 \cdot 2 = 1$ is correct.

Mathematical induction:
1) Show that $A(1)$ holds.
2) Show $A(n) \Rightarrow A(n + 1)$.
From 1) and 2) it follows:
$A(n)$ holds for all $n \in \mathbb{N}$.

$$S = 1 + \quad 2 \quad + \cdots + n$$
$$S = n + (n - 1) + \cdots + 1$$
Add the numbers on top of each other:
$$2S = \underbrace{(n + 1) + (n + 1) + \cdots + (n + 1)}_{n\,\text{times}}$$
$$2S = n(n + 1) \Rightarrow S = \tfrac{1}{2}n(n + 1)$$

# What does it mean for a sequence to converge to zero?

We say that a sequence converges to zero if the absolute value of the terms of the sequence becomes arbitrarily small for sufficiently large $n$.

Arbitrarily small means that you can specify any number greater than zero, usually called $\epsilon$, and from a certain index onward, the sequence only has terms whose absolute value is less than $\epsilon$.

The sequence $a_n = (-1)^n \dfrac{2}{n}$ converges to zero:
The terms $a_n$ become arbitrarily small for $n \to \infty$.

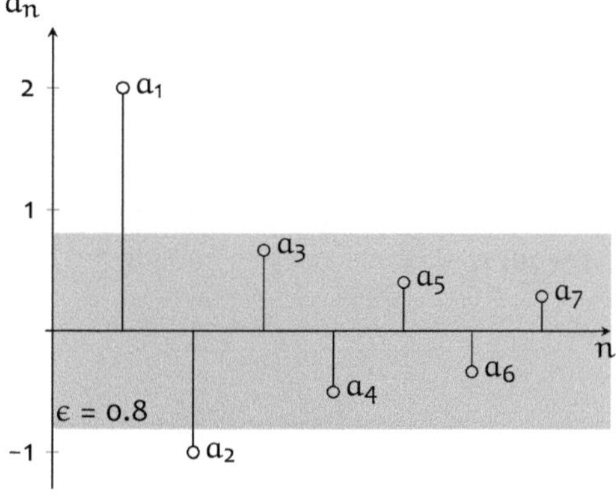


For any arbitrary $\epsilon > 0$ there is an N such that for $n > N$ all sequence terms lie in a strip of width $\epsilon$ around the x-axis.

For $\epsilon = 0.8$, all $a_n$ with $n > 2$ lie in the strip.

For smaller $\epsilon$, the strip becomes narrower and the required N becomes larger.

What matters is that for any $\epsilon > 0$ there is a corresponding N.

If the sequence converges to zero, you can also say that the limit of the sequence is zero and write this as a limit.

$$\lim_{n \to \infty} \frac{1}{n} = 0$$

The convergence of a sequence makes a statement about the behavior at infinity. It does not depend on finitely many terms.

$a_n = \frac{2}{n}$ for $n \geq 1$ and

$$b_n = \begin{cases} n^2 & \text{for } 1 \leq n \leq 1000 \\ \frac{2}{n} & \text{for } n \geq 1001 \end{cases}$$

have the same convergence behavior and the same limit.

# How do you use the definition to check whether a sequence converges to zero?

A sequence $(a_n)$ converges to zero if for every $\epsilon > 0$ there exists an N such that for all $n > N$ the absolute value $|a_n|$ is less than $\epsilon$.

To check this, you choose an $\epsilon$ with $\epsilon > 0$ and check whether there is an index N such that the absolute value of all terms $(a_n)$ with $n > N$ are less than $\epsilon$, i.e. whether $|a_n| < \epsilon$ for all $n > N$.

This must work for every $\epsilon > 0$. In other words, it must hold for all $\epsilon > 0$.

If this is the case, the sequence converges to zero.

Often, it's not easy to prove with the definition that a sequence converges. Therefore, limit theorems are helpful – which are calculation rules for limits.

For any $\epsilon > 0$ there exists an N such that $|a_n| < \epsilon$ for all $n > N$.
It is necessary to assume $\epsilon > 0$, because otherwise the absolute value can't be less than $\epsilon$.

For the sequence $a_n = \dfrac{1}{n}$:

If $\epsilon = 0.1$, then for $n > 10 \, (= N)$ the absolute value

$|a_n| = \dfrac{1}{n}$ is less than $\epsilon$:

$$n > 10 \Rightarrow \frac{1}{n} < \frac{1}{10} = 0.1$$

This works similarly for $\epsilon = 0.01$, the corresponding N is then 100.

For every $\epsilon > 0$ you can find an N (which depends on $\epsilon$), such that for $n > N$ we have $|a_n| < \epsilon$:

Choose $N = \dfrac{1}{\epsilon}$. Then for $n > N$:

$$|a_n| = \frac{1}{n} < \frac{1}{N} < \epsilon.$$

Note that when taking the reciprocal, the inequality sign reverses.

$$\lim_{n \to \infty} \frac{1}{n} = 0$$

# What does it mean for a sequence to converge to a limit?

If a sequence converges to 0, this means that for every $\epsilon > 0$ there exists an N such that $|a_n| < \epsilon$ for all $n > N$.

Convergence to a number other than zero is defined similarly: if all terms from a certain index onward lie within a strip around $\lim_{n\to\infty} a_n = a$, we say that the sequence converges to $a$.

There must be, for every $\epsilon > 0$, an N such that for all $n > N$, all $a_n$ lie within the highlighted strip.

Then, we say that the sequence converges to $a$. The number $a$ is called the limit.

When a sequence converges, it means that there is a real number (i.e. not $\infty$ or $-\infty$) to which the sequence converges.

$\lim\limits_{n\to\infty} a_n = 0$ holds if and only if for all $\epsilon > 0$ there exists an N such that for all $n > N$: $|a_n| < \epsilon$.

$\lim\limits_{n\to\infty} a_n = a$ holds if and only if for all $\epsilon > 0$ there exists an N such that for all $n > N$: $|a_n - a| < \epsilon$.

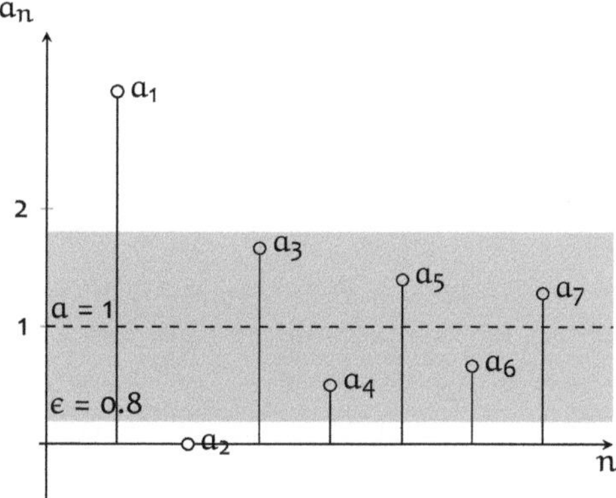


For $\epsilon = 0.8$, all $a_n$ with $n > 2$ lie within the strip.

The sequence converges to 1:
$$\lim_{n\to\infty} \frac{n + 2 \cdot (-1)^n}{n} = 1.$$

If $(a_n)$ converges, we write

$\lim\limits_{n\to\infty} a_n = a$, or, in short,

$a_n \to a \quad (n \to \infty)$.

# What does it mean for a sequence to diverge?

If there is no $a \in \mathbb{R}$ to which the sequence converges, then the sequence is called divergent, you can also say, it diverges.

These sequences diverge:
$$a_n = (-1)^n : \quad (a_n) = (-1, 1, -1, 1, \ldots)$$
$$b_n = n : \quad (b_n) = (1, 2, 3, 4, \ldots)$$
$$c_n = (-1)^n n : \quad (c_n) = (-1, 2, -3, 4, \ldots)$$

If the sequence grows beyond any bound, this can also be represented by the limit being infinity ($\infty$). Note: $\infty$ is not a real number.

$$\lim_{n \to \infty} n = \infty$$

Similarly, if the sequence falls below any negative number, this can also be represented by the limit being minus infinity ($-\infty$).

$$\lim_{n \to \infty} (-n) = -\infty$$

In these two cases the sequence is called properly divergent.

It can also happen that a sequence does not converge because it keeps oscillating between two values, or because it alternately becomes larger than any positive number and smaller than any negative number. There are also many other possibilities.

The sequences
$(a_n) = (-1, 1, -1, 1, \ldots)$ and
$(c_n) = ((-1)^n n) = (-1, 2, -3, 4, \ldots)$
are not properly divergent.

# How do you determine whether a sequence converges?

You can check the convergence of a sequence using the definition. However, other methods are often easier.

Above all, it's useful to know when a sequence of powers converges. This is exactly the case if the exponent is less than or equal to zero.

A negative exponential function converges to zero for any base b > 1.

Moreover, it approaches zero faster than any power function.

If two sequences converge, then the sum, difference, product, and quotient (as long as the denominator does not converge to zero) also converge.

With these rules, you can already determine the limits of many sequences.

Sometimes you first need to perform transformations.

If, in the case of a rational function, the degree of the denominator is greater than the degree of the numerator, then the limit as $n \to \infty$ is 0.

To do this, you must specify, for any $\epsilon > 0$, an N such that $|a_n - a| < \epsilon$ holds for all $n > N$.

$$\lim_{n \to \infty} n^{\alpha} = \begin{cases} 0 & \text{for} & \alpha < 0 \\ 1 & \text{for} & \alpha = 0 \\ \infty & \text{for} & \alpha > 0 \end{cases}$$

$$\lim_{n \to \infty} b^{-n} = 0 \quad \text{for } b > 1$$

$$\lim_{n \to \infty} b^{-n} n^{\alpha} = 0 \quad \text{for } b > 1, \alpha \in \mathbb{R}$$

With $\lim_{n \to \infty} a_n = a, \lim_{n \to \infty} b_n = b$:

$$\lim_{n \to \infty} (a_n \pm b_n) = a \pm b$$

$$\lim_{n \to \infty} (a_n \cdot b_n) = a \cdot b$$

$$\lim_{n \to \infty} \frac{a_n}{b_n} = \frac{a}{b}, \quad \text{if } b \neq 0.$$

$$\lim_{n \to \infty} n^2 \, 2^{-n} = 0$$

$$\lim_{n \to \infty} \frac{3n + 17}{5n + 1} = \lim_{n \to \infty} \frac{3 + \frac{17}{n}}{5 + \frac{1}{n}} = \frac{3}{5}$$

$$\lim_{n \to \infty} \frac{50n^2 + 3n + 100}{n^3 + 7} =$$

$$= \lim_{n \to \infty} \frac{\frac{50}{n} + \frac{3}{n^2} + \frac{100}{n^3}}{1 + \frac{7}{n^3}} = \frac{0}{1} = 0$$

# Why do monotone and bounded sequences converge?

Intuitively, a sequence can diverge in two ways: Either it grows beyond all bounds (positively or negatively), or it oscillates forever without approaching a limit.

It holds that if a sequence increases monotonically (then it does not oscillate) and is bounded above (then it cannot grow beyond all bounds), it converges. Similarly, a monotonically decreasing sequence that is bounded below, converges.

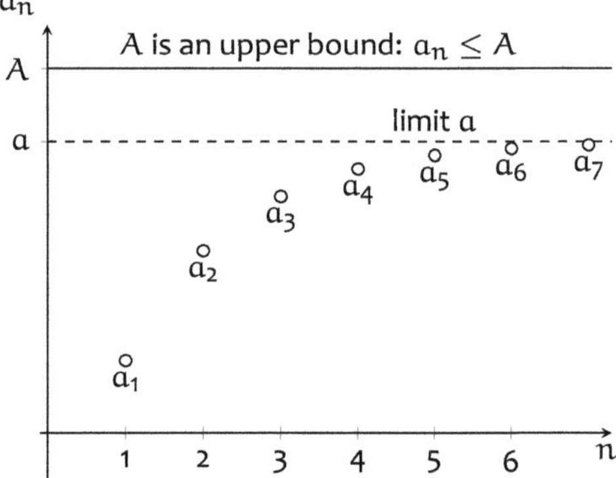


It's also sufficient if both statements hold only from a certain N onwards, i.e. for all $n > N$.

This is due to the completeness of the set of real numbers: the set of real numbers has no holes.

This does not apply to rational numbers: there are sequences of rational numbers that are monotonic and bounded but do not converge to a rational number, for example the sequence of decimal numbers with $n$ decimal places of $\sqrt{2}$ or $\pi$.

With this criterion, one can prove the convergence of a sequence without knowing its limit.

It can be shown that $a_n = \left(1 + \frac{1}{n}\right)^n$ is bounded by 3 (by multiplying out the parentheses using the binomial theorem and estimating the binomial coefficients from above) and that the sequence $(a_n)$ grows monotonically. From this, it follows that the sequence converges. The limit is Euler's number $e = 2.7182818\ldots$.

# What is a series and when does it converge?

Intuitively, a series is an infinite sum.

$$S = \frac{1}{2} + \frac{1}{4} + \frac{1}{8} + \ldots = \sum_{n=1}^{\infty} \frac{1}{2^n}$$

Mathematically, a series is a sequence whose terms $s_n$ are the sum of the first $n$ terms, also called partial sums. We say that the series converges if the sequence of partial sums converges.

$$s_1 = \frac{1}{2}$$

$$s_2 = \frac{1}{2} + \frac{1}{4}$$

$$\vdots$$

$$s_n = \frac{1}{2} + \frac{1}{4} + \ldots + \frac{1}{2^n}$$

A necessary criterion for the convergence of $\sum a_n$ is that the sequence $(a_n)$ converges to zero; in other words, $(a_n)$ must be a null sequence.

This can be seen as follows:
If a series $s_n = \sum_{i=1}^{n} a_i$ converges to $s$, then $\lim_{n\to\infty} a_n = \lim_{n\to\infty}(s_n - s_{n-1}) = \lim_{n\to\infty} s_n - \lim_{n\to\infty} s_{n-1} = s - s = 0$.

However, this criterion is not sufficient.

$$\text{For } \sum_{n=1}^{\infty} \frac{1}{n} \text{ we have } \lim_{n\to\infty} \frac{1}{n} = 0,$$

but the series diverges.

You can see this if you collect terms consecutively in groups of 2, 4, 8, ... as shown on the right. The sum of the terms in each group is greater than 1/2, and since there are infinitely many groups, the sum can't converge. The series on the right is also called the harmonic series.

$$1 + \frac{1}{2} + \underbrace{\frac{1}{3} + \frac{1}{4}}_{\geq 2\cdot\frac{1}{4}=\frac{1}{2}} +$$

$$+ \underbrace{\frac{1}{5} + \frac{1}{6} + \frac{1}{7} + \frac{1}{8}}_{\geq 4\cdot\frac{1}{8}=\frac{1}{2}} +$$

$$+ \underbrace{\frac{1}{9} + \frac{1}{10} + \frac{1}{11} + \frac{1}{12} + \frac{1}{13} + \frac{1}{14} + \frac{1}{15} + \frac{1}{16}}_{\geq 8\cdot\frac{1}{16}=\frac{1}{2}} +$$

$$+ \ldots \longrightarrow \infty$$

For an alternating series, the signs alternate between + and −. If the sequence forms a null sequence, the alternating series converges.

$$S = 1 - \frac{1}{2} + \frac{1}{3} - \frac{1}{4} \pm = \sum_{n=1}^{\infty} \frac{(-1)^{n+1}}{n}$$

converges, since it is alternating and

$$\lim_{n\to\infty} \frac{(-1)^{n+1}}{n} = 0.$$

# How do you determine the value of a geometric series?

In a geometric series, each term is multiplied by a constant factor to produce the next term.

You can easily calculate the value S of the series by observing that $qS$ gives the same series except for the first term, i.e. 1.

If you subtract the second equation from the first, you get an equation from which you can directly determine the value $S$ of the series.

With this equation you can determine the sum of the series for e.g. $q = 1/2$.

The geometric series converges for $0 < q < 1$ (and for $-1 < q \leq 0$).

The sum of the geometric series can also be derived geometrically, most simply by the intersection of the lines $y = qx$ and $y = x - 1$. In the figure, $q = 1/2$.

$$S = 1 + q + q^2 + \ldots = \sum_{n=0}^{\infty} q^n$$

$$qS = q + q^2 + \ldots = \sum_{n=0}^{\infty} q^{n+1}$$

$$(1 - q)S = 1$$
$$\Rightarrow S = \frac{1}{1 - q}$$

$$1 + \frac{1}{2} + \frac{1}{4} + \frac{1}{8} + \ldots = \frac{1}{1/2} = 2$$

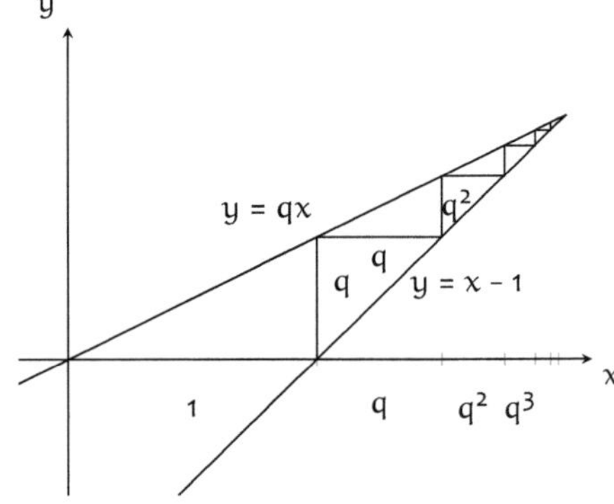


$$x = 1 + q + q^2 + q^3 + \ldots$$
$$qx = x - 1$$
$$\Rightarrow x = \frac{1}{1 - q}$$

# How can you determine the convergence of a series?

For the geometric series, the ratio of two consecutive terms is constant, as is the $n^{\text{th}}$ root of the n-th term (if the series starts with 1).

$$a_0 = 1, a_1 = q, a_2 = q^2, \ldots, a_n = q^n$$
$$\frac{a_{n+1}}{a_n} = q, \quad \sqrt[n]{a_n} = q$$

$$S = \sum_{n=0}^{\infty} a_n \text{ converges for } |q| < 1.$$

If the ratio or the root is not constant, but for $n \to \infty$ approaches a limit the absolute value of which is less than 1, then the corresponding series still converges.

$$\lim_{n \to \infty} \left| \frac{a_{n+1}}{a_n} \right| < 1$$
$$\Rightarrow \lim_{n \to \infty} \sum_{i=0}^{n} a_i \text{ converges.}$$

$$\lim_{n \to \infty} \sqrt[n]{|a_n|} < 1$$
$$\Rightarrow \lim_{n \to \infty} \sum_{i=0}^{n} a_i \text{ converges.}$$

With this ratio or root criterion, you can determine the convergence of many series.

$$\text{For } a_n = \frac{10^n}{n!} \text{ we have:}$$

$$\frac{a_{n+1}}{a_n} = \frac{10^{n+1}}{(n+1)!} \frac{n!}{10^n} = \frac{10}{n} \to 0$$

$$\Rightarrow \sum_{n=0}^{\infty} \frac{10^n}{n!} \text{ converges.}$$

# What are limits of functions?

The limit of a function can be defined using the limit of sequences.

We say, a function $f(x)$ has the limit $g$ as $x \to a$ if for every sequence $(x_n)$ converging to $a$, the sequence $(f(x_n))$ converges to $g$.

This also applies to the limit of a function $f(x)$ as $x \to \infty$. The limit of a function $f(x)$ as $x \to \infty$ is defined as the limit of the function values $f(x_n)$, where $x_n$ is a sequence with improper limit $\infty$, and the limits of the function values are the same for all such sequences.

If $(f(x_n))$ for some sequences $(x_n) \to \infty$ has one limit and for other sequences $(x_n) \to \infty$ another one, then the limit of the function $f(x)$ does not exist as $x \to \infty$.

$\lim\limits_{x \to a} f(x) = g$ means:
for every sequence $(x_n)$
with $\lim\limits_{n \to \infty} x_n = a$,
it holds that $\lim\limits_{n \to \infty} f(x_n) = g$.

For example, $\lim\limits_{x \to a} x^2 = a^2$, since
$$\lim\limits_{x \to a} (x \cdot x) = \left( \lim\limits_{x \to a} x \right) \cdot \left( \lim\limits_{x \to a} x \right) = a \cdot a.$$

$\lim\limits_{x \to \infty} \dfrac{1}{x} = 0$, since $\lim\limits_{n \to \infty} \dfrac{1}{x_n} = 0$ for
every sequence $(x_n) \to \infty$, e.g. for
$x_n = n$ or $x_n = n^2$ or $x_n = \sqrt{n}$.

$\lim\limits_{x \to \infty} \sin(\pi x)$ does not exist:

For $x_n = n$ with $\lim\limits_{n \to \infty} x_n = \infty$, we
get $\sin(\pi x_n) = 0$, so $\lim\limits_{n \to \infty} \sin(\pi x_n) = 0$.

For $x_n = \dfrac{4n+1}{2}$ with $\lim\limits_{n \to \infty} x_n = \infty$ we
get $\sin(\pi x_n) = 1$, so $\lim\limits_{n \to \infty} \sin(\pi x_n) = 1$.

Thus, for $\sin(\pi x)$ the limit $x \to \infty$ does not exist. This matches our intuition.

# How do you calculate simple limits?

Many limits can be reduced to known limits. Above all, it's interesting to know how a power function behaves in the limit $x \to \infty$: it approaches zero if and only if the exponent is negative.

$$\lim_{x \to \infty} x^\alpha = \begin{cases} 0 & \text{for} & \alpha < 0 \\ 1 & \text{for} & \alpha = 0 \\ \infty & \text{for} & \alpha > 0 \end{cases}$$

For $x \to 0$, it's the other way around: the power function approaches zero if and only if the exponent is positive.

$$\lim_{x \to 0} x^\alpha = \begin{cases} \infty & \text{for} & \alpha < 0 \\ 1 & \text{for} & \alpha = 0 \\ 0 & \text{for} & \alpha > 0 \end{cases}$$

A negative exponential function, $b^{-x}$, approaches zero for every $b > 1$ as $x \to \infty$.

$$\lim_{x \to \infty} b^{-x} = 0 \text{ for } b > 1$$

Moreover, it approaches zero faster than any power function.

$$\lim_{x \to \infty} b^{-x} x^\alpha = 0 \text{ for } b > 1, \alpha \in \mathbb{R}$$

The limit laws carry over from sequences to limits of functions.

With $\lim_{x \to x_0} f(x) = a$, $\lim_{x \to x_0} g(x) = b$:

$$\lim_{x \to x_0} (f(x) \pm g(x)) = a \pm b$$

$$\lim_{x \to x_0} (f(x) \cdot g(x)) = a \cdot b$$

$$\lim_{x \to x_0} \frac{f(x)}{g(x)} = \frac{a}{b}, \quad \text{if } b \neq 0.$$

With these rules, you can already evaluate many limits.

Sometimes you first have to perform some rearrangements.

$$\lim_{x \to \infty} \frac{3x + 17}{5x + 1} = \lim_{x \to \infty} \frac{3 + \frac{17}{x}}{5 + \frac{1}{x}} = \frac{3}{5}$$

If, in the case of a rational function, the degree of the denominator is greater than the degree of the numerator, then the limit for $x \to \infty$ is 0.

$$\lim_{x \to \infty} \frac{50x^2 + 3x + 100}{x^3 + 7} =$$

$$= \lim_{x \to \infty} \frac{\frac{50}{x} + \frac{3}{x^2} + \frac{100}{x^3}}{1 + \frac{7}{x^3}} = \frac{0}{1} = 0$$

# What does it mean intuitively for a function to be continuous?

A function is continuous at a point $x_0$ if small changes in the $x$-value only lead to small changes in the function value $f(x)$. The function must be defined at the point $x_0$ in the first place, which we will assume from now on.

When you want to find the right temperature of water in the shower and you tilt the knob then you want the temperature (let's say $y$) to depend continuously on how much you tilt (let's say $x$).

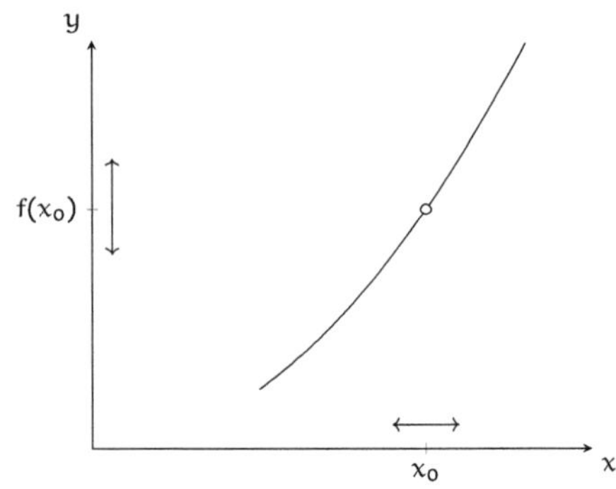

$f(x)$ is continuous: small changes in $x$ lead to small changes in $y$.

If the function has a jump at some point, it is not continuous.

$$f(x) = \begin{cases} 0 & \text{for } x < 0 \\ 1 & \text{for } x \geq 0 \end{cases}$$

is not continuous at $x = 0$:
If you move $x$ a little bit (no matter how little) around $x = 0$, $y$ changes drastically.

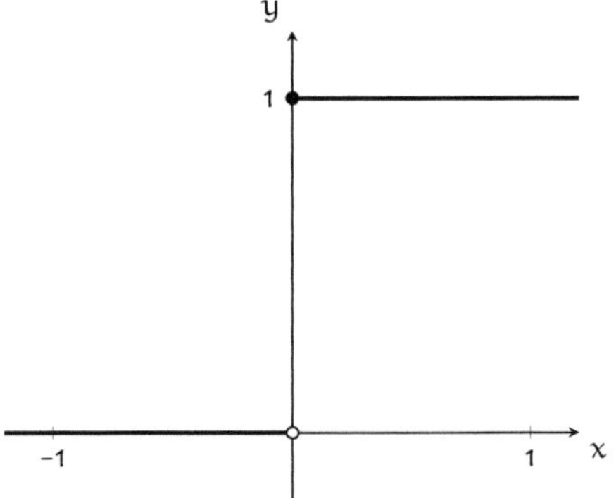

If the function oscillates so strongly at one point that it does not have a limit, it is not continuous.

$$f(x) = \begin{cases} \sin(\tfrac{1}{x}) & \text{for } x \neq 0 \\ 0 & \text{for } x = 0 \end{cases}$$

is not continuous at $x = 0$.

If you move $x$ just a little, $y$ fluctuates a lot.

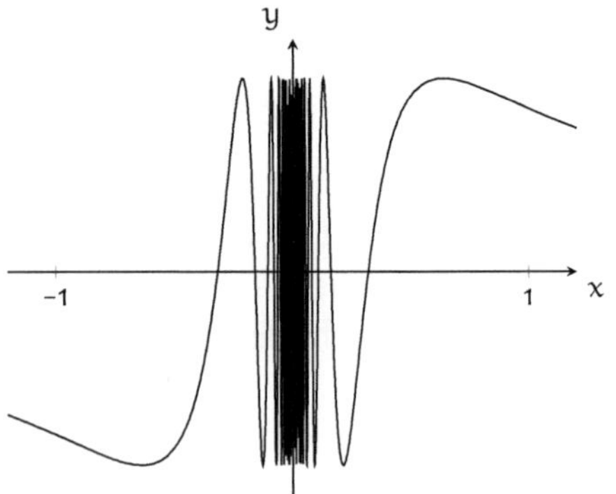

# What does it mean mathematically for a function to be continuous?

In order for a function to be continuous, you must be able to find, for every arbitrarily small neighborhood of the y-value a neighborhood of the x-value such that all function values of $x$ in the neighborhood of $\hat{x}$ lie within the neighborhood of $\hat{y}$.

It can be shown that this is equivalent to the following: For every sequence $(x_n)$ that converges to the x-value $\hat{x}$, the function values $f(x_n)$ converge to the corresponding y-value $f(\hat{x})$.

A function $f$ is continuous at $\hat{x}$, if for every $\epsilon > 0$ there is a $\delta > 0$ such that for all $x$ with $|x - \hat{x}| < \delta$ $|f(x) - f(\hat{x})| < \epsilon$ holds.

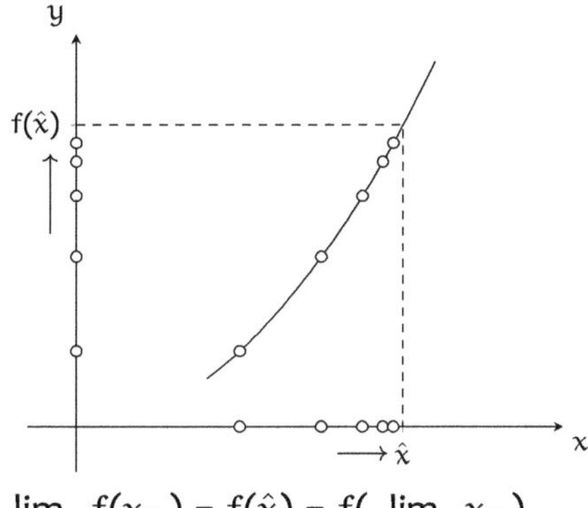


$$\lim_{n \to \infty} f(x_n) = f(\hat{x}) = f(\lim_{n \to \infty} x_n)$$

Thus it does not matter whether you first take the function value $f$ and then the limit $n \to \infty$ (left hand side) or first take the limit $n \to \infty$ and then the function value (right right hand side).

That the order does not matter can also be represented as a diagram. If both paths from left up to right down yield the same result, the function is continuous.

$$
\begin{array}{ccc}
x_n & \xrightarrow{\ \lim\ } & \hat{x} \\
f \downarrow & & f \downarrow \\
f(x_n) & \xrightarrow{\ \lim\ } & \begin{array}{l} f(\hat{x}) = \\ \lim_{n \to \infty} f(x_n) \end{array}
\end{array}
$$

$f$ is continuous at $x = \hat{x}$.

For a function to be continuous at a point $\hat{x}$, for all sequences $(x_n)$ with $\lim_{n \to \infty} x_n = \hat{x}$, the limits $\lim_{n \to \infty} f(x_n)$ must exist and be the same. And its value must be equal to the function value at that point.

$\iff$

For every sequence $(x_n) \to \hat{x}$:
a) $\lim_{n \to \infty} f(x_n)$ exists,
b) these limits are equal,
c) its value equals $f(\hat{x})$.

# What does it mean for a function to have a continuous extension?

If a function is not defined at a certain point, the question of continuity at that point can't be meaningfully asked.

In this case, one can only ask whether the function has a continuous extension, i.e. whether it's possible to assign a function value to the x-value where it is not defined such that the extended function is continuous.

However, there are functions that do have a continuous extension.

By the continuous extension, the hole is closed. This example may look a bit artificial; you could say that the denominator $x - 1$ was deliberately introduced to create a hole in the domain.

Something similar occurs in a very natural way, e.g. with the function $\sin(x)/x$, which is initially not defined at $x = 0$, but has a continuous extension by setting $f(0) = 1$. This function describes the diffraction at a slit in optics.

$$f(x) = \frac{1}{x} \text{ for } x \neq 0$$

is not defined at $x = 0$.

$$f(x) = \frac{1}{x} \text{ for } x \neq 0$$

doesn't have a continuous extension at $x = 0$.
No matter how you define $f(0)$, the resulting function defined for all $x \in \mathbb{R}$ would not be continuous.

$$f(x) = \frac{x^2 - 1}{x - 1} \text{ for } x \neq 1 \text{ has a}$$

continuous extension by setting

$$f(1) = \lim_{x \to 1} f(x) = \lim_{x \to 1} \frac{x^2 - 1}{x - 1} =$$
$$= \lim_{x \to 1} \frac{(x - 1)(x + 1)}{x - 1} = \lim_{x \to 1}(x + 1) = 2.$$

The continuous extension of the function is $g(x) = x + 1, x \in \mathbb{R}$. Only the point $(1, 2)$ was missing, since the original function was not defined at $x = 1$.

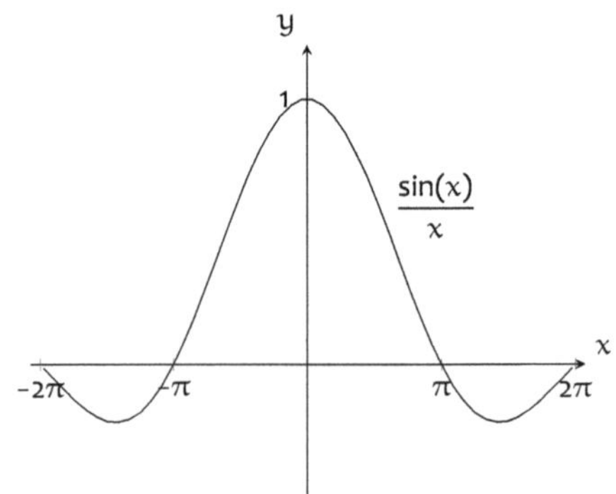

# How do you determine whether a function is continuous?

All functions that we have covered so far are continuous on their entire domain.

These functions are continuous:

$$x^n, n \in \mathbb{Z} \qquad \text{for all } x \in \mathbb{R},$$
$$\sqrt{x} \qquad \text{for all } x \geq 0,$$
$$a^x, a \geq 0 \qquad \text{for all } x \in \mathbb{R},$$
$$\ln(x) \qquad \text{for all } x > 0,$$
$$\sin(x), \cos(x) \qquad \text{for all } x \in \mathbb{R}.$$

Also sums, differences, products, quotients, and compositions of continuous functions are continuous on the domain of the resulting function.

These functions are continuous:

$$f(x) = x + \sin(x) \qquad \text{for } x \in \mathbb{R},$$
$$f(x) = \sqrt{x^2 + 1} \qquad \text{for } x \in \mathbb{R},$$
$$f(x) = \ln(x) + \ln(1 - x) \quad \text{for } x \in (0, 1).$$

The domain can be smaller than the original domain, specifically for quotients of functions, as the domain must exclude the zeros of the function in the denominator.

Not defined are

$$f(x) = \frac{x^2 - 1}{x - 1} \quad \text{for } x = 1,$$
$$f(x) = \frac{x - 1}{x^2 - 1} \quad \text{for } x = \pm 1.$$

Sometimes, it's possible to find a continuous extension of the function, in particular if a term cancels out.

$$f(x) = \frac{x - 1}{x^2 - 1}, x \neq \pm 1 \text{ can be}$$
continuously extended at $x = 1$ to the function
$$g(x) = \frac{1}{x + 1}, x \neq -1,$$
but not at $x = -1$.

The easiest way to prove continuity is by reducing the function to a known continuous function (see above).

If you have piecewise functions, proceed as described on the next page.

# How do you check a piecewise function for continuity?

A function is continuous on its entire domain if it is continuous at every point in its domain.

If you want to check the continuity of a piecewise function, you must first check whether the individual pieces are continuous, and then whether the function is continuous at the matching endpoints of the individual pieces.

You check this by taking the left-hand limit from the left function and the right-hand limit from the right function, which is equal to the function value if the function value is still part of the domain. If both match and are equal to the function value at that point, then the function is continuous.

It's possible that $x = a$ is included in the domain of the first function, then you need to check the limit of the second function.

$$f(x) = \begin{cases} g(x) & \text{for } x < a \\ h(x) & \text{for } x \geq a \end{cases}$$

is continuous, if
a) $g(x)$ is continuous for $x < a$,
b) $h(x)$ is continuous for $x \geq a$,
c) $\lim\limits_{x \to a} g(x) = h(a)$.

$$f(x) = \begin{cases} x + 2 & \text{for } x < 1 \\ 4 - x & \text{for } x \geq 1 \end{cases}$$

The two individual functions are continuous as linear functions.

Furthermore, $\lim\limits_{x \to 1}(x + 2) = 3 = 4 - 1$.

Therefore, $f(x)$ is continuous for all $x \in \mathbb{R}$.

$$f(x) = \begin{cases} x + 2 & \text{for } x \leq 1 \\ 4 - x & \text{for } x > 1 \end{cases}$$

The two individual functions are continuous as linear functions.

Furthermore, $\lim\limits_{x \to 1}(4 - x) = 3 = 1 + 2$.

Therefore, $f(x)$ is continuous for all $x \in \mathbb{R}$.

# Differential Calculus

We now come to what is called calculus, which was founded in the $17^{\text{th}}$ century by Leibniz and Newton.

With the help of differential calculus, you can calculate the slope of the tangent to a curve at a point. This is the limit of the slope of a line through two points of the curve as the second point moves towards the first and unifies.

The value of the slope of the graph of a function is the derivative of the function at a given point. This new function describes the rate of change of the value of the original function at each $x$-value.

The derivative of the position of a body with respect to time similarly gives the change of position per unit time, i.e. the velocity of the body. Here, the limit allows the transition from average velocity to instantaneous velocity.

By means of differential calculus, i.e. the calculation of the derivative, special points of a curve such as maxima and minima can be determined efficiently.



A. Gründers, *Math Made Clear: From the Basics to Calculus*, https://doi.org/10.1007/978-3-662-73221-2_9

# How do you determine the slope of the tangent at a point on a curve?

The tangent at a point $P_0$ on a curve is obtained when you consider the line through this point $P_0$ and a neighboring point $P$, and then move $P$ ever closer to $P_0$ until the two points meet. As a simple case, we consider the standard parabola $f(x) = x^2$ and the point $P_0(1, 1)$.

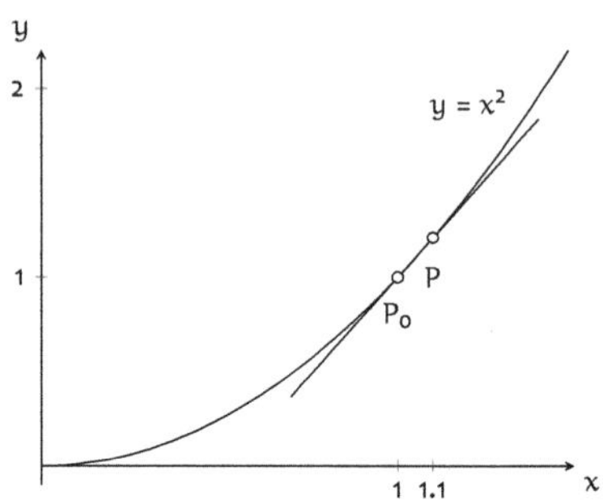


As the $x$-value of the point $P$, we choose $x = 1 + h$.

$x_P = 1 + h$ (above, $h = 0.1$)
The smaller the value of $h$, the closer $P$ is to $P_0$.

The $y$-value of the point $P$ is obtained by substituting $x$ into the function equation.

$$y_P = x_P^2 = (1 + h)^2$$

The slope of the line through $P_0(1, 1)$ and the point $P(1 + h, (1 + h)^2)$ is calculated from the ratio of the difference of the $y$-values to the difference of the $x$-values.

$$\frac{\Delta y}{\Delta x} = \frac{y_P - y_0}{x_P - x_0} = \frac{(1 + h)^2 - 1}{(1 + h) - 1}$$

$$= \frac{1 + 2h + h^2 - 1}{h} = \frac{2h + h^2}{h} = 2 + h$$

For the limit $h \to 0$, we obtain the slope of the tangent at $x = 1$.

$$\lim_{h \to 0} (2 + h) = 2$$

We write $f'(x)$ for the slope of the tangent at the point $P(x, f(x))$, and call $f'(x)$ the derivative.

$$f'(1) = \lim_{h \to 0} \frac{f(1 + h) - f(1)}{h} = 2$$

Similarly to $x = 1$ you can calculate the derivative at any point $x$.

$$f'(x) = \lim_{h \to 0} \frac{f(x + h) - f(x)}{h}$$

$$= \lim_{h \to 0} \frac{(x + h)^2 - x^2}{h}$$

$$= \lim_{h \to 0} \frac{x^2 + 2xh + h^2 - x^2}{h}$$

$$= \lim_{h \to 0} \frac{2xh + h^2}{h} = \lim_{h \to 0} (2x + h) = 2x$$

# What does the derivative mean?

The derivative of a function $f(x)$ at a point $x$ gives the slope of the tangent to the graph of the function at the point $P(x, f(x))$.

$f'(x)$ = slope of the tangent at $x$

If $f(x) = x^2$,
then $f'(x) = 2x$, $f'(1) = 2$.

The slope of the tangent is the local rate of change of the function value with respect to the independent variable.

$y = f(x) = x^2$ changes at $x = 1$
with the rate of change $f'(1) = 2$:
If $x$ changes from 1 by 0.01 to 1.01,
then $x^2$ changes by $\approx f'(1) \cdot 0.01 = 0.02$.

For the local rate of change $f'(x)$, we also write $dy/dx$. The "d" stands for differential.

$$\frac{dy}{dx} = f'(x)$$

The unit of measure of the derivative (noted in square brackets) is given by the ratio of the unit of measure of the function value to the unit of measure of the independent variable.

$$\left[\frac{dy}{dx}\right] = \frac{[dy]}{[dx]} = \frac{[y]}{[x]}$$

For the slope of the curve, the unit of measure of $y$ and $x$ is the same, so the derivative is dimensionless.

$$[y] = cm, [x] = cm, \frac{[dy]}{[dx]} = \frac{cm}{cm} = 1$$

If the function describes the position $s(t)$ of an object as a function of time, the derivative $s'(t) = v(t)$ gives the velocity. The unit of measure is m/s.

$$[s] = m, [t] = s, \frac{[ds]}{[dt]} = \frac{m}{s} = [v]$$

If you are given $s(t)$ with correct units, so that substituting $t$ in seconds yields a length in meters, then the derivative will automatically have the correct unit of measure, i.e. meters per second.

$$s(t) = 100\,m - 5\frac{m}{s^2}t^2$$

$$v(t) = s'(t) = -10\frac{m}{s^2}t$$

$$v(1\,s) = s'(1\,s) = -10\frac{m}{s}$$

# When is a function differentiable?

A function is differentiable at a point $x$ in its domain if the derivative, i.e. the differential quotient as the limit of the difference quotient does exist.

For this to be the case, the limit of the difference quotient $\dfrac{f(x + h) - f(x)}{h}$ must exist for every sequence $h \to 0$.

From differentiability follows continuity, but not vice versa. There are continuous functions that are not differentiable, such as $f(x) = |x|$ at $x = 0$. The graph of the function $f(x) = |x|$ has a corner at $x = 0$, the derivative does not exist.

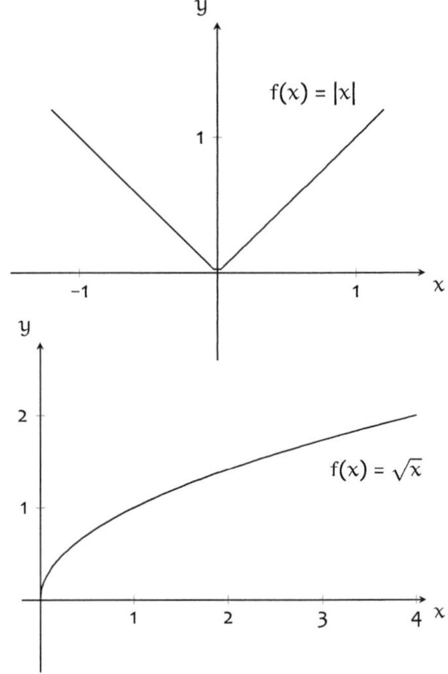


The function $f(x) = \sqrt{x}$ is not differentiable at $x = 0$, since the limit of the differential quotient does not exist. We have

$$f'(0) = \lim_{h \to 0} \frac{\sqrt{h} - 0}{h} = \infty.$$

There are also more complicated examples of non-differentiability than just a simple corner or the derivative being $\infty$.

$$f(x) = \begin{cases} x \sin(\tfrac{1}{x}) & \text{for } x \neq 0 \\ 0 & \text{for } x = 0 \end{cases}$$

is continuous at $x = 0$, but not differentiable.

Equivalent to the definition via the existence of the difference quotient is local linear approximability.

$f(x)$ is differentiable at $x = x_0$ if for $h \to 0$ there exists a number $m \in \mathbb{R}$ such that $f(x_0 + h) = f(x_0) + m\,h + r(h)$, where the remainder $r(h)$ goes to zero faster than linearly: $r(h)/h \to 0$ as $h \to 0$.

# How do you differentiate power functions?

Similarly to the derivative of $y = x^2$, you can also calculate the derivative of $y = x^3$.

$$y = f(x) = x^3$$

$$f'(x) = \lim_{h \to 0} \frac{(x + h)^3 - x^3}{h}$$

$$= \lim_{h \to 0} \frac{x^3 + 3x^2 h + 3xh^2 + h^3 - x^3}{h}$$

$$= \lim_{h \to 0} \frac{3x^2 h + 3xh^2 + h^3}{h}$$

$$= \lim_{h \to 0} (3x^2 + 3xh + h^2)$$

$$= 3x^2$$

This can be generalized to any $n \in \mathbb{N}$ and even holds for arbitrary real exponents.

$$y(x) = x^n$$
$$y'(x) = nx^{n-1}$$

Two important examples arise for the exponents −1 and 1/2.

$$y = f(x) = x^{-1} = \frac{1}{x}$$

$$y' = f'(x) = -1 \cdot x^{-1-1} = -x^{-2} = -\frac{1}{x^2}$$

$$y(x) = x^{1/2} = \sqrt{x}$$

$$y'(x) = \frac{1}{2} x^{1/2 - 1} = \frac{1}{2} x^{-1/2} = \frac{1}{2\sqrt{x}}$$

This differentiation rule holds for all exponents.

$$f(x) = \sqrt[3]{x} = x^{\frac{1}{3}}$$

$$f'(x) = \frac{1}{3} x^{\frac{1}{3} - 1} = \frac{1}{3} x^{-\frac{2}{3}} =$$

$$= \frac{1}{3x^{\frac{2}{3}}} = \frac{1}{3\sqrt[3]{x^2}}$$

$$f(x) = x^{-0.45}$$
$$f'(x) = -0.45 \, x^{-1.45}$$

Note that as the derivative of a power function, you can never get an exponent of −1, because for that to be the case $n - 1 = -1$ would have to hold, i.e. $n = 0$, but then $nx^{n-1} = 0$.

When we later reverse differentiation and look for functions that have a given derivative, this will be important.

# How do you differentiate sums of functions?

The derivative of the sum or difference of functions is simply the sum or difference of the derivatives.

$$f(x) = x^2 \Rightarrow f'(x) = 2x$$
$$g(x) = x^3 \Rightarrow f'(x) = 3x^2$$
$$h(x) = x^2 + x^3$$
$$\Rightarrow h'(x) = 2x + 3x^2$$

In general:
$$((f(x) + g(x))' = f'(x) + g'(x)$$

If a function is multiplied by a constant, simply differentiate the function and multiply by the constant.

$$f(x) = x^2 \Rightarrow f'(x) = 2x$$
$$f(x) = 3x^2 \Rightarrow f'(x) = 6x$$

In general: $(af(x))' = af'(x)$

This is because you can pull the factor in front of the limit when taking the limit.

$$(af(x))' = \lim_{h \to 0} \frac{af(x + h) - af(x)}{h} =$$
$$= a \lim_{h \to 0} \frac{f(x + h) - f(x)}{h} = af'(x)$$

It's very important to note this only works if the factor is a constant, i.e. a fixed number. If two functions are multiplied together, you need the product rule, which we will get to later.

If you combine both rules, you can differentiate functions of the form $af(x) \pm bg(x)$.

$$(af(x) + bg(x))' = af'(x) + bg'(x)$$

Now you can differentiate any polynomial. The derivative of a polynomial of degree $n$ is a polynomial of degree $n - 1$.

$$p(x) = \sum_{i=0}^{n} a_i x^i = a_0 + a_1 x + \cdots + a_n x^n$$
$$\Rightarrow p'(x) = \sum_{i=0}^{n} i a_i x^{i-1} =$$
$$= a_1 + 2a_2 x + \cdots + n x^{n-1}$$

# What is the slope of the exponential function at $x = 0$?

The slope of an exponential function at any point can be reduced to the slope at the point $x = 0$.

For $f(x) = a^x$:

$$f'(x) = \lim_{h \to 0} \frac{f(x + h) - f(x)}{h} =$$

$$\lim_{h \to 0} \frac{a^{x+h} - a^x}{h} = \lim_{h \to 0} \frac{a^x a^h - a^x}{h}$$

$$= a^x \lim_{h \to 0} \frac{a^h - 1}{h} = a^x f'(0)$$

Let us now consider the slope at $x = 0$.

$$f'(0) = \lim_{h \to 0} \frac{a^h - 1}{h}$$

By calculating this limit for $a = 2$ numerically, you find a number that is approximately 0.6931472.

$$\frac{2^{0.1} - 1}{0.1} \approx 0.718, \qquad \frac{2^{0.01} - 1}{0.01} \approx 0.696$$

$$\lim_{h \to 0} \frac{2^h - 1}{h} = 0.6931472 \ldots$$

For $a = 3$, you similarly get a number that is a bit greater than 1.

$$\lim_{h \to 0} \frac{3^h - 1}{h} = 1.0986123 \ldots$$

The graphs of the functions $2^x$ and $3^x$ thus have at $x = 0$ a slope that is less than or greater than 1, respectively.

It's a reasonable conjecture, which can be proved, that there is a unique number between 2 and 3 for which the slope is exactly 1.

We call this number Euler's number e. It is approximately 2.718281828.

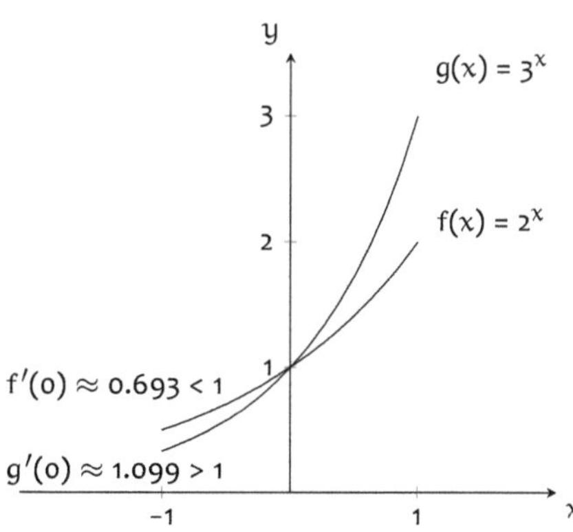

# What is special about the exponential function with slope 1 at $x = 0$?

The base, for which the exponential function has slope 1 at $x = 0$, is called e. It is defined by the corresponding limit.

$$\lim_{h \to 0} \frac{e^h - 1}{h} = 1$$

We solve for e by formally rearranging the expression inside the limit. The proof of correctness has been omitted.

$$e = \lim_{h \to 0} (1 + h)^{1/h}$$

Now we set $n = 1/h$ and thus, instead of $h \to 0$, we consider the limit $n \to \infty$.

$$e = \lim_{n \to \infty} \left(1 + \frac{1}{n}\right)^n$$

The limit of this sequence is e.

$$\left(1 + \frac{1}{2}\right)^2 = 2.25$$

$$\left(1 + \frac{1}{10}\right)^{10} = 2.5937\ldots$$

$$\left(1 + \frac{1}{100}\right)^{100} = 2.7048\ldots$$

$$e = \lim_{n \to \infty} \left(1 + \frac{1}{n}\right)^n = 2.71828\ldots$$

The function has slope 1 at $x = 0$, and even more, it has (see the top of the previous page) at every point a slope equal to its function value.

$$f(x) = e^x \Rightarrow f'(0) = 1$$

$$f'(x) = e^x f'(0) = e^x f'(0) = f(x)$$

The function is its own derivative! It can be shown that, up to a multiplicative constant, it's the only function that is its own derivative.

$$y'(x) = y(x)$$

$$\Rightarrow y(x) = ce^x, c \in \mathbb{R}$$

# How else can the $e^x$ function be written?

The defining equation for e can also be written as a series using the binomial theorem. We will not show this in detail.

$$e = \lim_{n \to \infty} \left(1 + \frac{1}{n}\right)^n$$

$$\text{with } (1 + x)^n = \sum_{k=0}^{n} \binom{n}{k} x^k, \; x = \frac{1}{n}$$

and $n \to \infty$ gives

$$e = 1 + \frac{1}{1!} + \frac{1}{2!} + \frac{1}{3!} + \ldots = \sum_{k=0}^{\infty} \frac{1}{k!}$$

For $e^x$ you can proceed similarly. To do this, we first rewrite $e^x$.

$$e^x = \lim_{n \to \infty} \left(1 + \frac{1}{n}\right)^{nx}$$

Set $m = nx$, $n = \frac{m}{x}$, then as $n \to \infty$, $m \to \infty$ as well, and it follows

$$e^x = \lim_{m \to \infty} \left(1 + \frac{x}{m}\right)^m.$$

If you then reapply the binomial theorem as above, you arrive at the series representation of the exponential function.

$$f(x) = e^x = 1 + \frac{x}{1!} + \frac{x^2}{2!} + \frac{x^3}{3!} + \frac{x^4}{4!} + \ldots$$

$$= \sum_{n=0}^{\infty} \frac{x^n}{n!}$$

Also in this representation, you can see that the derivative of the function, $f(x) = e^x$, is itself. Each term in the series, when differentiated, gives the previous one. Since there are infinitely many terms, you get exactly the same expression.

$$f(x) = e^x = 1 + \frac{x}{1!} + \frac{x^2}{2!} + \frac{x^3}{3!} + \frac{x^4}{4!} + \ldots$$

$$f'(x) = 1 + \frac{2x}{2!} + \frac{3x^2}{3!} + \frac{4x^3}{4!} + \ldots$$

$$= 1 + \frac{x}{1!} + \frac{x^2}{2!} + \frac{x^3}{3!} + \ldots$$

$$= e^x$$

$$= f(x)$$

# In what sense is the derivative a linearization?

The tangent is a straight line. So, if you approximate a function close to a point by its tangent at that point, you have linearized it.

The linearization is an approximation of the function near a point. It only takes into account the function value at the point and the linear change depending on the difference $h$ of the $x$-values. Quadratic terms (proportional to $h^2$) and further nonlinear terms are neglected.

These terms, which go to 0 faster than $h$ as $h \to 0$, are denoted by $o(h)$.

These higher terms become very small for a small $h = x - 1$, i.e. close to the considered $x$-value, 1.

For small $h$, the linearization is therefore a good approximation for differentiable functions.

The coefficient of $h$ in the linearization is precisely the derivative, i.e. the slope of the tangent, which indicates how much the function changes near that point in the linear approximation.

This also holds in general.

$f(x) = x^2$ near $x = 1$:
Set $x = 1 + h$ with small $h = x - 1$:

$$f(1 + h) = (1 + h)^2 = \underbrace{1 + 2h}_{\text{linearization}} + h^2$$

$$f(x) = f(1 + h) \approx \underbrace{1}_{\substack{\text{func-}\\\text{tion}\\\text{value}}} + \underbrace{2h}_{\substack{\text{linear}\\\text{term}}}$$

$$f(x) = f(1 + h) = 1 + 2h + o(h)$$

If $h = 0.1$ then
$h^2 = 0.01$ and
$h^3 = 0.001$.

$(1 + h)^2 \approx 1 + 2h$
$1.1^2 \approx 1.2$    (exactly: 1.21)
$1.01^2 \approx 1.02$    (exactly: 1.0201)

$f(x) = x^2$, for $x = 1$:
$$(1 + h)^2 = \underbrace{1}_{f(1)} + \underbrace{2}_{f'(1)}\, h + o(h)$$

For a general point $x_0$:
$$(x_0 + h)^2 = \underbrace{x_0^2}_{f(x_0)} + \underbrace{2x_0}_{f'(x_0)}\, h + o(h)$$
with $f'(x_0) = 2x_0$, i.e. $f'(x) = 2x$.

$$f(x + h) = f(x) + hf'(x) + o(h)$$

# How can you use the fact that the derivative is the linearization?

If you can give a linear approximation of a function, you can read off the derivative directly.

$$f(x) = x^3$$
$$(x_0 + h)^3 = \underbrace{x_0^3}_{f(x_0)} + \underbrace{3x_0^2}_{f'(x_0)} h + o(h)$$

Thus: $f'(x_0) = 3x_0^2$, i.e. $f'(x) = 3x^2$

The variable $h$ stands for the deviation of the $x$-values from the point under consideration ($x_0 = 1$). Instead of $h$, you can also write $\Delta x = x - x_0$ or $dx$.

$dx$ is called the differential and it can be regarded as a linear map (which leads to differential forms) or as an infinitesimal number (not a real number, but a hyperreal number). Both is out of scope here.

If you write $dx$, you omit the higher order terms.

$$(x + dx)^3 = x^3 + 3x^2 dx$$

This works for any differentiable function as well.

$$f(x + dx) = f(x) + f'(x)dx$$

You can then also write the change in $y$ as $dy$.

$$y + dy = f(x + dx) = \underbrace{f(x)}_{y} + \underbrace{f'(x)dx}_{dy}$$

If you formally divide by $dx$, you obtain a new notation for the derivative, which sometimes has advantages, but should be used with caution. It is not a fraction in the usual sense.

$$dy = f'(x)\,dx$$
$$\frac{dy}{dx} = f'(x)$$

# How do you differentiate a product?

The derivative of the product of two functions is the sum of the derivative of the first function times the second and the first function times the derivative of the second.

$$(uv)' = u'v + uv'$$

$$f(x) = \underbrace{x^2}_{u}\ \underbrace{e^x}_{v}$$

i.e. $u = x^2$, $v = e^x$,
thus $u' = 2x$, $v' = e^x$,

$$f'(x) = \underbrace{2x}_{u'}\ \underbrace{e^x}_{v} + \underbrace{x^2}_{u}\ \underbrace{e^x}_{v'}.$$

You can understand this by considering the change in the product of the two functions up to the first order in h. In the third line, we write $u$ instead of $u(x)$, etc.

$$u(x + h)v(x + h)$$

$$= (u(x) + hu'(x))(v(x) + hv'(x)) + o(h)$$

$$= uv + \underbrace{(u'v + uv')}_{(uv)'}h + \underbrace{u'v'h^2}_{o(h)}$$

You can also understand this by interpreting the product, $u(x)v(x)$, as an area and considering how the area changes in linear approximation in h.

| $hv'$ | $huv'$ | $\propto h^2$ |
|---|---|---|
| $v(x)$ | $uv$ | $hu'v$ |

$$u(x) \qquad hu'$$

linear change in area: $(u'v + uv')h$

You can also write the product rule using differentials.

$$d(uv) = udv + vdu$$

$$u = x^2,\ du = 2xdx$$
$$v = e^x,\ dv = e^x dx$$

$$d(x^2 e^x) = x^2 e^x dx + e^x 2x dx =$$
$$= (x^2 + 2x)e^x dx$$

# How do you differentiate a quotient?

The derivative of the quotient of two functions is the derivate of the numerator times the denominator minus the numerator times the derivative of the denominator, and all this divided by the denominator squared.

$$\left(\frac{u}{v}\right)' = \frac{u'v - uv'}{v^2}$$

$$y(x) = \frac{x}{1 + x^2}$$

i.e. $u = x, v = 1 + x^2$,
thus $u' = 1, v' = 2x,$

$$y'(x) = \frac{1 \cdot (1 + x^2) - 2x \cdot x}{(1 + x^2)^2}$$

$$= \frac{1 - x^2}{(1 + x^2)^2}.$$

This quotient rule can be derived in a similar way as the product rule.

You cleverly expand the fraction that you obtain by $v - hv'$, so that there is no longer a term linear in $h$ in the denominator, but only a quadratic term $h^2$. You can neglect this quadratic term for determining the derivative as only the linear change in $h$ counts. From the second line on, we write $u$ instead of $u(x)$, etc.

$$\frac{u(x + h)}{v(x + h)} = \frac{u(x) + hu'(x)}{v(x) + hv'(x)}$$

$$= \frac{(u + hu'))(v - hv')}{(v + hv')(v - hv')}$$

$$= \frac{uv + h(u'v - uv') - h^2u'v'}{v^2 - h^2v'^2}$$

$$= \frac{uv + h(u'v - uv')}{v^2} + o(h)$$

$$= \frac{u}{v} + h \underbrace{\frac{u'v - uv'}{v^2}}_{\left(\frac{u}{v}\right)'} + o(h)$$

The linear term is the derivative and gives you the quotient rule.

$$\left(\frac{u}{v}\right)' = \frac{u'v - uv'}{v^2}$$

# What is a composite function?

For a composite function, two functions are nested within each other: you first calculate one function, then apply the second function to its value ($z$ for intermediate result).

The function you calculate first is called the inner function (since it often appears inside, in parentheses), and the function you calculate second is called the outer function.

N.B. The function you calculate first appears second (further to the right) when reading from left to right.

In general, the order matters.

Whenever under a root or as the argument of a function there is not just a simple $x$, but an expression in $x$, i.e. a function of $x$, it is a composite function.

Composite functions often occur in applications, e.g. when the force on a body depends on the position of the body, but the position in turn depends on time.

First calculate $x^2 (= z)$ from $x$, then sine: $\sin(z) = \sin(x^2)$.

You then get $y(x) = \sin(x^2)$. $\sin(x^2)$ is a composite function.

For $y(x) = \sin(x^2)$:
inner function: $g(x) = x^2$
outer function: $f(z) = \sin(z) = \sin(x^2)$

$y(x) = f(g(x))$
g: inner function
f: outer function

For $f(g(x))$, you first calculate $g(x)$, then f of $g(x)$.

$$\sin(x^2) \neq (\sin(x))^2$$

Composite functions:

$\sqrt{x + 1}$    inner function: $x + 1$

$\ln(2x)$    inner function: $2x$

$\dfrac{1}{x^2 + 1}$    inner function: $x^2 + 1$

$F(x(t))$: The force F depends on the position $x$, the position $x$ depends on the time t.

# How do you find the derivative of a composite function?

The derivative of a composite function is found by first differentiating the inner function and then multiplying by the derivative of the outer function (while simply keeping the inner function as is).

$$(\sin(x^2))' = \underbrace{2x}_{\substack{\text{derivative} \\ \text{of } x^2}} \cdot \underbrace{\cos}_{\substack{\text{derivative} \\ \text{of } \sin}} (x^2)$$

This is the chain rule: the derivative of a composite function is the inner derivative times the outer derivative.

$$f(g(x))' = g'(x)f'(g(x))$$

You can understand the chain rule by considering linearization. If you linearize $y(x) = f(g(x))$, you must first linearize the function $g$ at the point $x$ and then the function $f$ at the point $g(x)$. For the second step we don't have h, but $\widetilde{h} = hg'(x)$. If you read off the coefficient of h in the result, which is the derivative, you get the chain rule.

$$\begin{aligned}
f(g(x+h)) &= f(g(x)+hg'(x)+o(h)) = \\
&= f(g(x)+\widetilde{h})+o(h) \quad \text{with } \widetilde{h}=hg'(x) \\
&= f(g(x)) + \widetilde{h}f'(g(x)) + o(h) \\
&= f(g(x)) + h\underbrace{g'(x)f'(g(x))}_{(f(g(x)))'}+o(h)
\end{aligned}$$

You can also remember the chain rule using the $dy/dx$ notation.

$$y = f(g(x))$$
$$y' = \frac{dy}{dx} = \frac{d\,f(g(x))}{dx} = \frac{df}{dx} = \frac{df}{dg}\frac{dg}{dx}$$

If you have a multiple nested function, you work your way from the inside out and multiply the respective derivatives.

$$f(g(h(x)))' =$$
$$= h'(x)\,g'(h(x))\,f'(g(h(x)))$$
$$\left(\sin\left(e^{x^2}\right)\right)' = 2x\,e^{x^2}\cos\left(e^{x^2}\right)$$

# How do you differentiate exponential functions?

The exponential function with a general base $a^x$ can be rewritten as an exponential function with base e using the rules of exponentiation.

$$a^x = (e^{\ln(a)})^x = e^{x\ln(a)}$$

It is a composite function, whose derivative you can calculate using the chain rule.

$$f(x) = e^{x\ln(a)}$$

$$f'(x) = \ln(a)e^{x\ln(a)} = \ln(a)a^x$$

The slope of the exponential function $a^x$ at the point $x = 0$ is therefore the natural logarithm of a.

$$f'(0) = \ln(a)$$

Here you can see again that the function $e^x$ has slope 1 at $x = 0$, and all exponential functions $a^x$ with $a < e$ have a slope less than 1, and all with $a > e$ have a slope greater than 1.

$$f(x) = e^x \Rightarrow f'(0) = 1$$
$$f(x) = 2^x \Rightarrow f'(0) = 0.693\ldots < 1$$
$$f(x) = 3^x \Rightarrow f'(0) = 1.099\ldots > 1$$
see above

Earlier you already saw that the exponential function $e^x$ is special because it does not change when you differentiate it. It is often given its own name, $\exp(x)$.

$$\exp(x) = e^x$$
$$(\exp(x))' = \exp(x)$$

# How do you find the derivative of the inverse function?

The graph of the inverse function is the graph of the original function reflected across the straight line $y = x$.

This allows you to determine the derivative of the inverse function. I'll show you this using the example of the derivative of $\ln(x)$ at the point $x = e$.

The point $(e, 1)$ on the graph of the logarithm function corresponds to the point $(1, e)$ on the graph of the exponential function.

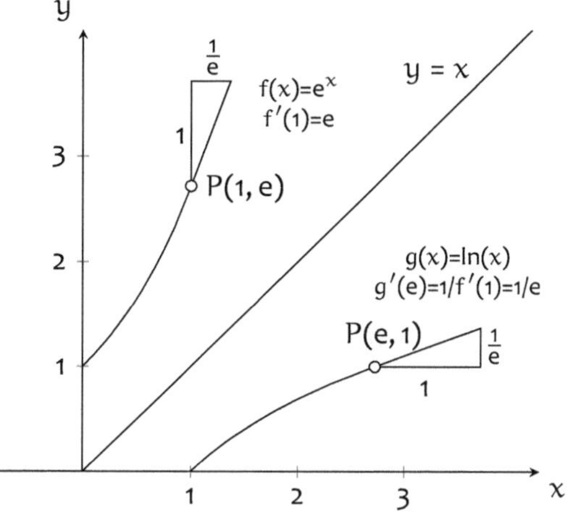


The corresponding slope triangles are also reflected across the line $x = y$, so the adjacent and opposite sides are swapped, and the slope of the second triangle is the reciprocal of the slope of the first.

$$g'(e) = \frac{1}{e} = \frac{1}{f'(1)}$$

This also holds at a general point.

$$g'(e^x) = \frac{1}{f'(x)}$$

$$\ln'(e^x) = \frac{1}{e^x} \Rightarrow \ln'(x) = \frac{1}{x}$$

Similarly, it also works for any other function and its inverse.

$g(x)$ is the inverse function of $f(x)$.

$$\Rightarrow g'(f(x)) = \frac{1}{f'(x)}$$

# How do you concretely determine the derivative of the inverse function?

You can also determine the derivative of the inverse function using the chain rule. For simplicity, we denote the inverse function by g.

In this way, you can determine the derivative of the logarithm function.

Similarly, you can differentiate the inverse function of the tangent function, called arctan(x).

Definition of the inverse function g:

$$g(f(x)) = x, \quad y = f(x), x = g(y)$$

Differentiate both sides:

$$f'(x)\, g'(f(x)) = 1$$

Solve for the derivative $g'$:

$$g'(f(x)) = \frac{1}{f'(x)}, \text{ or}$$

$$g'(y) = \frac{1}{f'(x)} \text{ or } g'(y) = \frac{1}{f'(g(y))}$$

The solution of $y = f(x) = e^x$ is $x = g(y) = \ln(y)$.

$$g'(y) = (\ln(y))' = \frac{1}{f'(x)} = \frac{1}{e^x} = \frac{1}{y}$$

$$\Rightarrow (\ln(x))' = \frac{1}{x}$$

The derivative of $\ln(-x), x < 0$ yields $1/x$ (chain rule), too. Therefore, $(\ln|x|)' = 1/x$.

The solution of $y = f(x) = \tan(x)$ is $x = g(y) = \arctan(y)$.

$$g'(y) = \frac{1}{f'(x)} = \frac{1}{\tan'(x)}$$

Derivative of $\tan(x) = \frac{\sin(x)}{\cos(x)}$:

$$\tan'(x) = \frac{\cos(x)\cos(x) - (-\sin(x)\sin(x))}{\cos^2(x)}$$

$$= 1 + \frac{\sin^2(x)}{\cos^2(x)} = 1 + \tan^2(x)$$

Thus:

$$(\arctan(y))' = \frac{1}{\tan'(x)} =$$

$$= \frac{1}{1 + \tan^2(x)} = \frac{1}{1 + y^2}$$

$$\Rightarrow (\arctan(x))' = \frac{1}{1 + x^2}$$

# Overview: Functions and their derivatives

Here you will find the most important simple functions with their derivatives and the differentiation rules for composite functions.

| $y = f(x)$ | $y' = f'(x)$ |
|:---:|:---:|
| $x^n$ | $nx^{n-1}$ |
| $e^x$ | $e^x$ |
| $\ln(x)$ | $\dfrac{1}{x}$ |
| $\sin(x)$ | $\cos(x)$ |
| $\cos(x)$ | $-\sin(x)$ |
| $af(x) + bg(x)$ | $af'(x) + bg'(x)$ |
| $f(x)g(x)$ | $f'(x)g(x) + f(x)g'(x)$ |
| $\dfrac{f(x)}{g(x)}$ | $\dfrac{f'(x)g(x) - f(x)g'(x)}{g^2(x)}$ |
| $f(g(x))$ | $g'(x)f(g(x))$ |

You do not need to memorize the following derivatives, instead, you can easily derive them yourself:

$$(a^x)' = \left(e^{x\ln(a)}\right)' = \ln(a)e^{x\ln(a)} = \ln(a)a^x$$

$$(\log_a(x))' = \left(\frac{1}{\ln(a)}\ln(x)\right)' = \frac{1}{\ln(a)x}$$

$$(\tan(x))' = \left(\frac{\sin(x)}{\cos(x)}\right)' = \frac{\cos^2(x) + \sin^2(x)}{\cos^2(x)} = \frac{1}{\cos^2(x)}$$

# How do you determine possible solutions for minima and maxima?

At a local minimum $x_{min}$ of a function, the function has a local minimum, i.e. there is a neighborhood of $x_{min}$ such that the function value is greater everywhere else in that neighborhood.

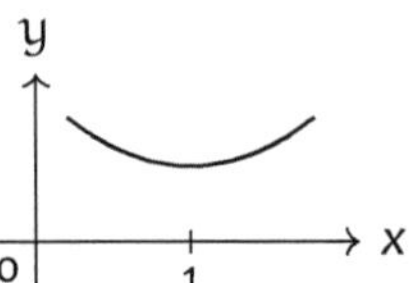

The function has a minimum at $x_{min} = 1$.

At a local maximum $x_{max}$ of a function, the function has a local maximum, i.e. there exists a neighborhood of $x_{max}$ such that the function value is smaller everywhere else in that neighborhood.

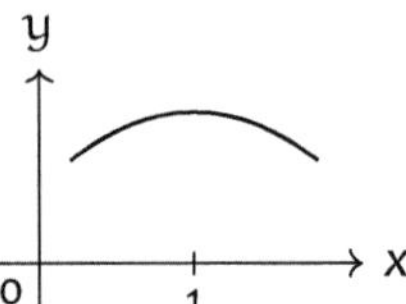

The function has a maximum at $x_{max} = 1$.

A necessary condition for a max or a min is that the derivative of the function at the respective point (if it exists) equals 0.

If the derivative at a maximum (min) were not 0, but greater than 0, then just to the right (left) of that point the function value would be greater (smaller), so there would not be a maximum (min) there. Similarly, if the derivative were less than 0, then the function value just to the left would be greater.

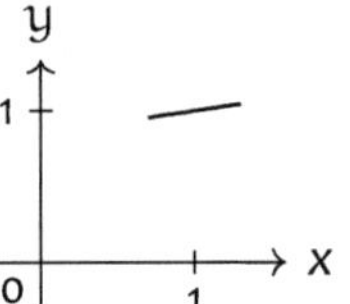

For $f'(1) > 0$, $x = 1$ can't be a maximum, since for small $\epsilon > 0$ it would hold that $f(1 + \epsilon) > f(1)$.

However, the condition $f'(x) = 0$ is not sufficient. It can also be the case that $f'(x) = 0$ without there being a maximum or minimum. This is called a saddle point.

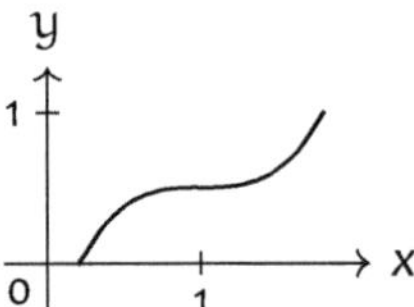

Here, $f'(1) = 0$, but $f(1)$ is not a min/max, but a saddle point.

164

# How do you determine whether a solution of $f'(x) = 0$ is a minimum or maximum?

If you go from left to right through a minimum, you first go down ($f' < 0$), at the very bottom it is flat ($f' = 0$), then you go up again ($f' > 0$).

$f'(x_0) = 0$
and $f'(x) < 0$ for $x < x_0$
and $f'(x) > 0$ for $x > x_0$
$\Rightarrow x_0$ is a minimum and $f(x_0)$ is a minimum value

If you go from left to right through a maximum, you first go up ($f' > 0$), at the very top it is flat ($f' = 0$), then you go down again ($f' < 0$).

$f'(x_0) = 0$
and $f'(x) > 0$ for $x < x_0$
and $f'(x) < 0$ for $x > x_0$
$\Rightarrow x_0$ is a maximum and $f(x_0)$ is a maximum value

Therefore, to find the possible extrema, we start by differentiating the function and determining the zero(s) of $f'(x)$.

$f(x) = x^3 - 3x$
$f'(x) = 3x^2 - 3 = 0$
$x_{1,2} = \pm 1$

You must then check how $f'(x)$ changes sign at those points.

At $x = -1$, $f'$ changes from + to -, so it's a maximum.
At $x = 1$, $f'$ changes from - to +, so it's a minimum.

Alternatively, you can also calculate the second derivative $f''$, which indicates the curvature.

If $f''$ is greater than 0, the curve is concave up, so there is a minimum; if it is less than 0, the curve is concave down, so there is a maximum.

$f'' > 0$, i.e. concave up
$\Rightarrow$ minimum

$f'' < 0$, i.e. concave down
$\Rightarrow$ maximum

If the second derivative is 0, you can either look at the sign change of the first derivative (see above) or consider higher derivatives.

If the first nonzero derivative is of even order, there is an extremum. If it is of odd order, there is a saddle point.

# What does the second derivative tell you?

The second derivative is simply the derivative of the first derivative (which depends on $x$, so it is also a function).

$$f(x) = x^2 - 2x + 2$$
$$f'(x) = 2x - 2$$
$$f''(x) = 2$$

So $f''(x)$ indicates how quickly the slope of the function $f$ increases or decreases at the point $x$.

If the second derivative is positive, that is, the slope increases as the $x$-value increases, then the graph of the function is concave up: if you move in the direction of increasing $x$-values, you have to turn the handlebars to the left.

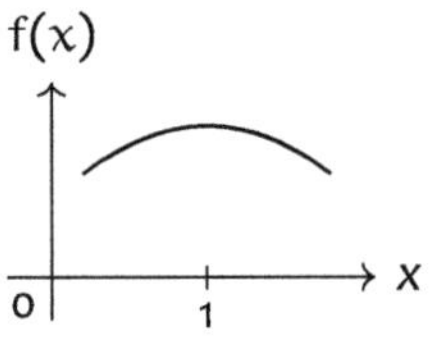


$f'' > 0$: concave up

If the second derivative is negative, that is, the slope decreases as the $x$-value increases, then the graph of the function is concave down: if you move in the direction of increasing $x$-values, you have to turn the handlebars to the right.



$f'' < 0$: concave down

If the function under consideration is the position as a function of time, then the first derivative is the rate of change of position with respect to time, i.e. the velocity.

Thus, the second derivative is the rate of change of velocity with respect to time, i.e. the acceleration.

When you accelerate, the acceleration is positive ($s''(t) > 0$), and the velocity increases.

When you decelerate, the acceleration is negative ($s''(t) < 0$), and the velocity decreases.

# What are inflection points, and how do you find them?

If the second derivative changes sign at a point $x$ (i.e. the graph changes from concave up to concave down, or vice versa), we say that there is an inflection point at $x$.

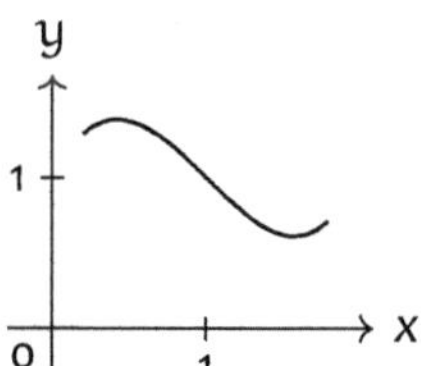

for $x < 1$: concave down
for $x > 1$: concave up
at $x = 1$: inflection point

It is similar to the first derivative. If the first derivative changes from positive to negative or from negative to positive, there is an extremum (maximum or minimum).

For inflection points, we do not distinguish between the two cases.

You find the inflection points by differentiating the function twice and finding the zeros of $f''(x)$.

$f(x) = x^3 - 3x^2 + 2x + 1$
$f'(x) = 3x^2 - 6x + 2$
$f''(x) = 6x - 6 = 0 \Rightarrow x = 1$

You then need to check whether $f''(x)$ changes sign at that point, or whether $f'''(x) \neq 0$. If $f'''(x) \neq 0$, you have found an inflection point. If $f'''(x) = 0$, an inflection point may or may not exist.

$f''(1) = 0$
$f'''(1) = 6 \neq 0$
$\Rightarrow$ inflection point at $x = 1$.

If the first nonzero derivative after $f''$ is of odd order, then there is an inflection point.

For $f(x) = x^5$ we have $f''(0) = 0$, $f'''(0) = 0$, $f^{(iv)} = 0$, $f^{(v)} = 120 \neq 0$.
$\Rightarrow$ inflection point at $x = 0$.

By the way, an inflection point with a horizontal tangent ($f'(x) = 0$) is called a saddle point.

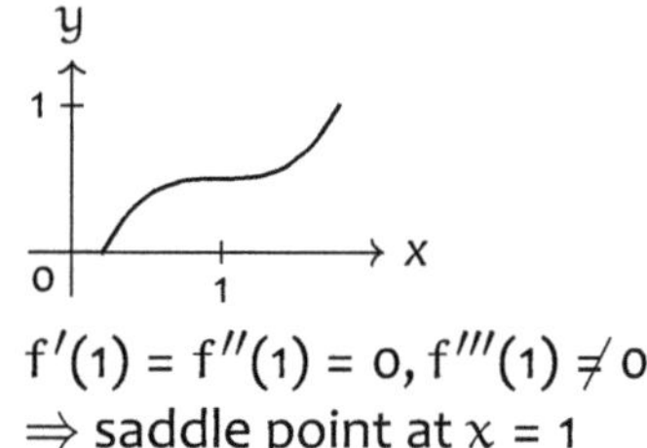

$f'(1) = f''(1) = 0$, $f'''(1) \neq 0$
$\Rightarrow$ saddle point at $x = 1$

# Why is the geometric mean less than or equal to the arithmetic mean? (I)

The arithmetic mean of two numbers is the number which, when added to itself, gives the sum of the two numbers, i.e. half the sum.

$$\overline{a}_{\text{arith}} + \overline{a}_{\text{arith}} = a_1 + a_2$$

$$\overline{a}_{\text{arith}} = \frac{1}{2}(a_1 + a_2)$$

The geometric mean of two numbers is the number which, when multiplied by itself, gives the product of the two numbers, i.e. the square root of the product.

$$\overline{a}_{\text{geom}} \cdot \overline{a}_{\text{geom}} = a_1 \cdot a_2$$

$$\overline{a}_{\text{geom}} = \sqrt{a_1 \cdot a_2}$$

Both definitions can be generalized for $n$ numbers.

$$\overline{a}_{\text{arith}} = \frac{1}{n}(a_1 + a_2 + \cdots + a_n)$$

$$\overline{a}_{\text{geom}} = \sqrt[n]{a_1 a_2 \cdots a_n}$$

For $n = 2$, one can easily prove that the arithmetic mean is at least as large as the geometric mean.

$$0 \le \left(\sqrt{a} - \sqrt{b}\right)^2$$

$$0 \le a - 2\sqrt{a}\sqrt{b} + b$$

$$\sqrt{ab} \le \frac{a + b}{2}$$

This inequality can also be recognized geometrically.

The two small right triangles with legs $a$ and $h$ or $h$ and $b$ are similar to each other, and thus $\frac{h}{a} = \frac{b}{h}$ holds, so the height of the large right triangle is $h = \sqrt{ab}$.

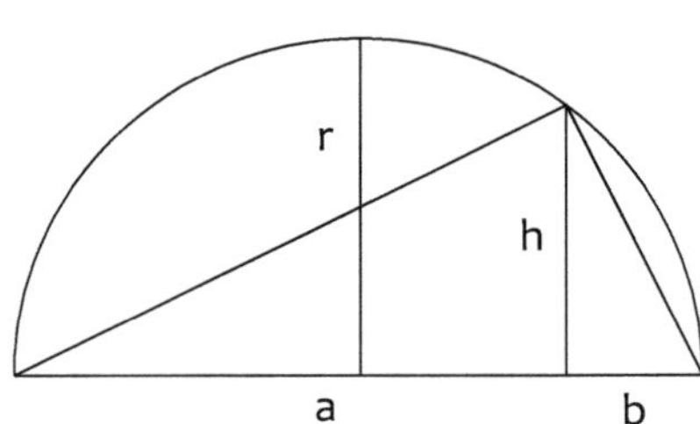


$$h = \sqrt{ab} \le \frac{a + b}{2} = r$$

The radius of the circle is $r = \frac{a+b}{2}$. Since the height of the triangle is less than or equal to the radius of the circle, it follows that $\sqrt{ab} \le \frac{a+b}{2}$.

# Why is the geometric mean less than or equal to the arithmetic mean? (II)

If you want to prove the inequality for $n > 2$, you can do so either by mathematical induction or by using an inequality from analysis, namely $e^x \geq 1 + x$.

$$e^x \geq 1 + x$$

Since the function $e^x$ is concave up (the second derivative $e^x$ is $> 0$ for all $x \in \mathbb{R}$), the graph of the function always lies above the tangent $y = 1 + x$ for $x \neq 0$.

For the proof, one estimates $\exp(a_i/A - 1)$ from below, where $A$ is the arithmetic mean.

$$\exp\left(\frac{a_i}{A} - 1\right) \geq 1 + \frac{a_i}{A} - 1 = \frac{a_i}{A}$$

Then one multiplies these inequalities for $i = 1, \ldots, n$ together.

$$\prod_{i=1}^{n} \exp\left(\frac{a_i}{A} - 1\right) \geq \prod_{i=1}^{n} \frac{a_i}{A}$$

The product of e to an exponent is e to the sum of the exponents, and on the right side the constant $A$, which is independent of $i$, can be factored out of the product.

$$\exp\left(\sum_{i=1}^{n}\left(\frac{a_i}{A} - 1\right)\right) \geq \frac{\prod_{i=1}^{n} a_i}{A^n}$$

The sum on the left hand side is zero because $\sum a_i = nA$. The product on the right hand side is, by the definition of the geometric mean, $G$, equal to $G^n$.

$$\exp(0) \geq \frac{G^n}{A^n}$$

Thus, it is proven that the geometric mean, $G$, is less than or equal to the arithmetic mean, $A$.

$$G \leq A$$

$$\sqrt[n]{a_1 a_2 \cdots a_n} \leq \frac{a_1 + a_2 + \cdots + a_n}{n}$$

If you replace $a_i$ by $1/a_i$, you find that the harmonic mean, $H$ (the reciprocal of the arithmetic mean of the reciprocals), is less than or equal to the geometric mean.

$$H = \frac{n}{\frac{1}{a_1} + \frac{1}{a_2} + \cdots + \frac{1}{a_n}}$$

$$H \leq G \leq A$$

# Overview: Determining extrema and inflection points

For functions that are differentiable arbitrarily many times, you determine extrema and inflection points in the interior of the domain as follows:

| Criteria for extrema |
| --- |
| necessary<br>$f'(x_0) = 0,\quad f''(x_0) > 0$ : minimum<br>$\qquad\qquad f''(x_0) < 0$ : maximum<br>$\qquad\qquad f''(x_0) = 0$ and $f'''(x_0) \neq 0$ : saddle point |
| necessary and sufficient<br>• $f'$ changes sign at $x_0$<br>$\quad$ from + to - : max;$\quad$ from - to + : min<br>• The first nonzero derivative $f^{(k)}$ is of even order k.<br>$\quad f^{(k)} > 0$ : min, $\quad f^{(k)} < 0$ : max<br>$\quad$ If $f'(x_0) = 0$ and k is odd: saddle point. |

| Criteria for inflection points |
| --- |
| necessary<br>$f''(x_0) = 0,\quad f'''(x_0) > 0$ : from concave down to concave up<br>$\qquad\qquad f'''(x_0) < 0$ : from concave up to concave down<br>$\qquad\qquad f'''(x_0) = 0$ and $f''''(x_0) \neq 0$ : extremum, see above. |
| necessary and sufficient<br>• $f''$ changes sign at $x_0$<br>$\quad$ from + to - : from concave up to down; from - to + : from concave down to up.<br>• The first nonzero derivative $f^{(k)}$ is of odd order k. |

# Overview: How are the graphs of f and f′ related?

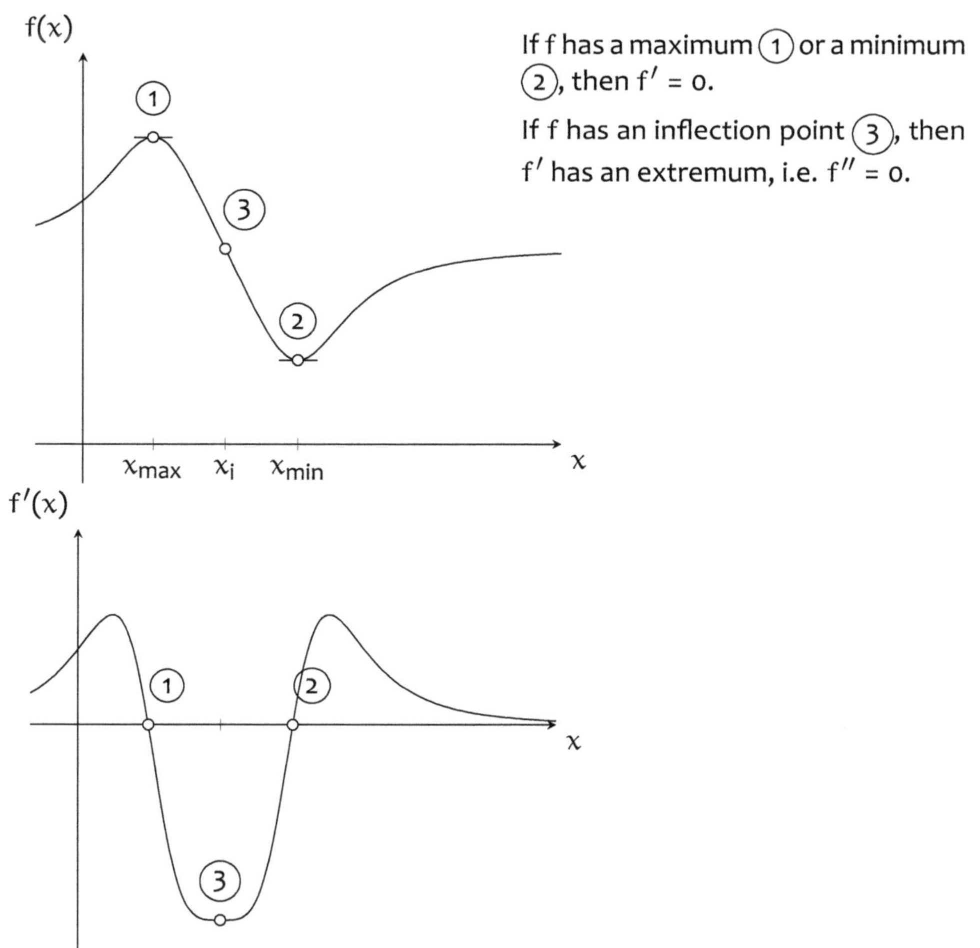



If f has a maximum ① or a minimum ②, then $f' = 0$.

If f has an inflection point ③, then $f'$ has an extremum, i.e. $f'' = 0$.

| f | increasing | decreasing | maximum | minimum | inflection point |
|---|---|---|---|---|---|
| f′ | > 0 | < 0 | = 0, decreasing | = 0, increasing | extremum |

# How do you solve optimization problems? (I)

An optimization problem is a word problem in which a quantity should be made as large (or as small) as possible.

■ If possible, make a sketch and label it.

■ Collect all quantities from the problem and the sketch, and name them. Consider in which units of measure you will specify them.

■ Consider which quantity should be maximized and set up an equation for it that shows which quantities it depends on.

■ If it depends on two (or more) quantities, consider how these quantities depend on each other; this is called a constraint.

■ Solve the constraint for one variable and substitute this variable into the quantity to be maximized.

■ The quantity to be maximized (let's call it f) then depends only on one independent variable (let's call it $x$).

How should the sides of a rectangle with a perimeter of 1 m be chosen so that the area is maximized?

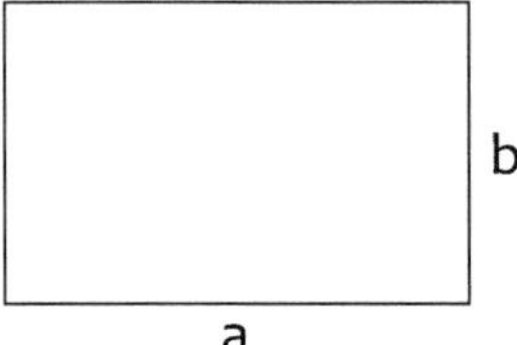

area (to be maximized): A
perimeter (given): P
side lengths (from sketch) a, b
lengths in m, areas in m²

A should be maximized.
$A = a \cdot b$

perimeter $P = 1$
$P = a + b + a + b = 2(a + b) = 1$

$$2(a + b) = 1 \Rightarrow b = \frac{1}{2} - a$$

$$A = ab = a\left(\frac{1}{2} - a\right)$$

$$f(x) = x\left(\frac{1}{2} - x\right) = \frac{1}{2}x - x^2$$

# How do you solve optimization problems? (II)

(continued from previous page)

- Then determine the maximum by setting $f'(x) = 0$, solving for $x$, and checking whether $f''(x) < 0$ (for a minimum $> 0$).

$$f'(x) = \frac{1}{2} - 2x = 0$$

$$x = \frac{1}{4}, \quad f''(x) = -2 < 0, \text{ ok}$$

- Translate the solution $x$ with the maximum value $f(x)$ back into the original problem and check for plausibility.

$$x = \frac{1}{4}, \quad f(x) = \frac{1}{4}\left(\frac{1}{2} - \frac{1}{4}\right) = \frac{1}{16}$$

$$a = \frac{1}{4}, \quad b = \frac{1}{2} - \frac{1}{4} = \frac{1}{4}, \quad A = \frac{1}{16}$$

$a$ and $b$ are equal, the area is positive. So this is ok.

- Formulate the answer as a complete sentence, and don't forget to include the units of measure.

The rectangle with the maximum area results when both sides are equal, i.e. when it's a square. Then they are $\frac{1}{4}$ m $= 25$ cm long, and the maximum area is $\frac{1}{16}$ m$^2$.

# How do you do a curve sketching?

We will show a curve sketching example using a fourth-degree polynomial.

$$y = f(x) = x^4 - x^2$$

It's best to write down the first three derivatives at the beginning.

$$f'(x) = 4x^3 - 2x$$
$$f''(x) = 12x^2 - 2$$
$$f'''(x) = 24x$$

First, you calculate the zeros of the function, if that's possible analytically. Plot these in a coordinate system (see below).

$$y = f(x) = x^4 - x^2 = 0$$
$$x^2(x^2 - 1) = 0$$
$$\Rightarrow a)\ x^2 = 0 \text{ or } b)\ x^2 - 1 = 0$$
a) $x_{1,2} = 0, \quad N_{1,2}(0,0)$ double zero.
b) $x_{3,4} = \pm 1, \quad N_3(-1,0), N_4(1,0)$

Then you calculate the zeros of the derivative, these are the candidates for the extrema, and test them with $f''(x)$. Mark these points with a small piece of horizontal tangent.

$$f'(x) = 0 \Rightarrow x(4x^2 - 2) = 0$$
a) $x = 0 \Rightarrow x_1 = 0$
$$f''(0) = -2 < 0 \Rightarrow \text{Max}(0,0)$$
b) $2x^2 - 1 = 0 \Rightarrow x_{2,3} = \pm\frac{1}{\sqrt{2}}$
$$f''\left(\pm\tfrac{1}{\sqrt{2}}\right) = 4 > 0 \Rightarrow \text{Min}\left(\pm\tfrac{1}{\sqrt{2}}, -\tfrac{1}{4}\right)$$

Afterwards, you calculate the zeros of the second derivative. These are the candidates for inflection points. Calculate the function value and the value of the derivative for these x-values and mark the points with a small piece of the tangent.

$$f''(x) = 0 \Rightarrow 12x^2 - 2 = 0$$
$$\Rightarrow x_{1,2} = \pm\frac{1}{\sqrt{6}}$$
$$f'''\left(\pm\tfrac{1}{\sqrt{6}}\right) \neq 0 \Rightarrow I_{1,2}\left(\pm\tfrac{1}{\sqrt{6}}, -\tfrac{5}{36}\right)$$

Finally, you connect the points and make sure that the curve follows at each point the direction of the respective tangent.

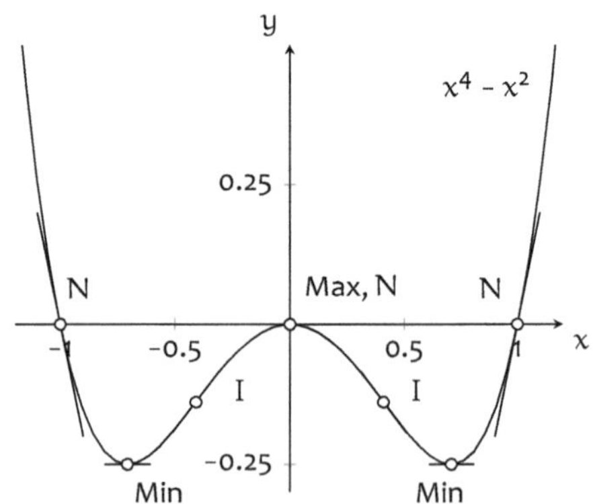

# Overview: Curve sketching

With a curve sketching, you get an overview of what the graph of a function looks like.

To do this, you collect as much information as possible about the function and its graph and enter the information graphically into a coordinate system (see example above).

| |
|---|
| Check whether there is an axial symmetry or a point symmetry. |
| Symmetry with respect to the y-axis: $f(-x) = f(x)$ <br> Point symmetry with respect to the origin: $f(-x) = -f(x)$ |
| Determine the axes intercepts |
| Determine the intersection with the y-axis: set $x = 0$ in $f(x)$. <br> Determine the zeros, i.e. set $f(x) = 0$ and solve for $x$. <br> Also calculate the derivative at these points. |
| Determine the extrema and inflection points. |
| Extrema: Set $f'(x) = 0$, solve for $x$ <br>  and check $f''(x) \neq 0$ or sign change of $f'(x)$. <br><br> Inflection points: Set $f''(x) = 0$, solve for $x$ <br>  and check $f'''(x) \neq 0$ or sign change of $f''(x)$. |
| Determine the behavior for $x \to \pm\infty$ and the asymptotes. |
| Sketch the graph. |

# Integral Calculus

With the help of integral calculus, it's possible to calculate areas under curves that are given as graphs of functions.

The crucial insight of Newton and Leibniz was that the rate of change of the area function is equal to the function that defines the graph.

Or, put a bit more loosely: the derivative of the area under a curve is the curve itself.

Similarly, integral calculus broadly allows the calculation of quantities from their rates of change.

For example, distance can be determined from velocity, velocity from acceleration, and the amount of liquid in a container from the inflow.

The generalization to higher dimensions is also straightforward: you can obtain the trajectory by integrating the velocity vector, and the velocity vector from the acceleration vector, i.e. the force.

It's possible to calculate volumes and surface areas of bodies bounded by curved surfaces, and much more.



A. Gründers, *Math Made Clear: From the Basics to Calculus*, https://doi.org/10.1007/978-3-662-73221-2_10

# How can you calculate areas using infinite sums?

To calculate the area under the parabola $y = x^2$ from $x = 0$ to $x = 1$, you can approximate it by dividing the area into $n$ narrow rectangles of width $h = 1/n$ and heights $0$, $h^2$, $(2h)^2$, …

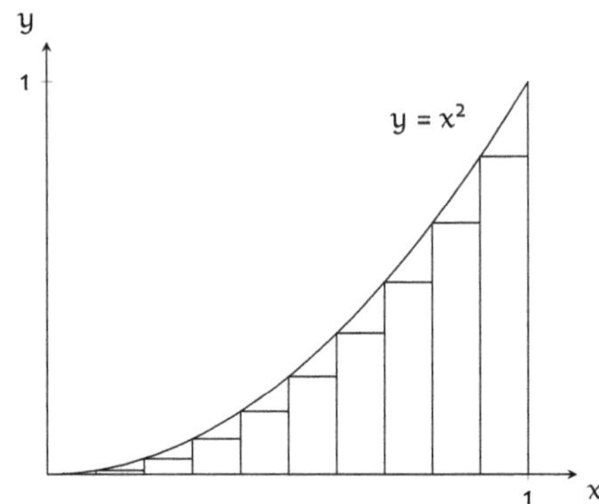



$n = 10$

The area under the parabola is obtained as the limit $n \to \infty$ of the area $A_n$: $A = \lim\limits_{n \to \infty} A_n$.

$$A_n =$$
$$= h\underbrace{\left(0^2 + h^2 + (2h)^2 + \ldots + ((n-1)h)^2\right)}_{n \text{ terms}}$$
$$= h^3 \sum_{k=0}^{n-1} k^2 \quad \text{with } h = \tfrac{1}{n}$$

To calculate you need the summation formula for the first $n$ squares, which you can prove using mathematical induction.

$$\sum_{k=0}^{n-1} k^2 = \frac{(n-1)\,n\,(2n-1)}{6}$$

With this, you find in the limit $n \to \infty$ the area under the parabola.

$$F = \lim_{n \to \infty} \frac{1}{n^3} \frac{(n-1)\,n\,(2n-1)}{6} = \frac{1}{3}$$

In this way, you can determine the area below the graphs of functions. Later, you will learn a much simpler method (keyword: antiderivative).

# What is a definite integral?

The definite integral over a positive function from $x = a$ to $x = b$ gives the area between the x-axis, the graph of the function, and the two vertical boundaries $x = a$ and $x = b$.

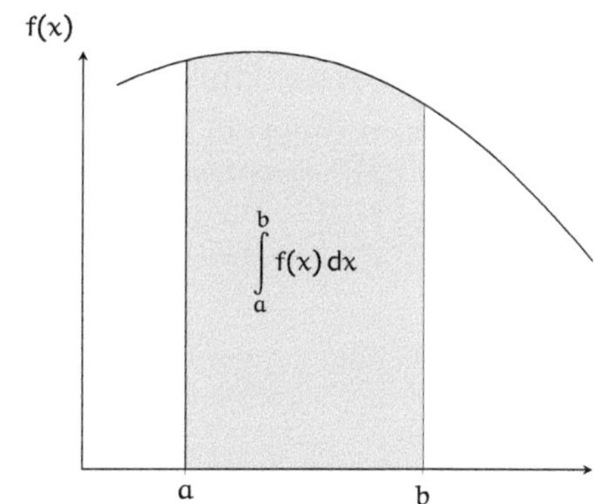



We call $a$ and $b$ the lower and upper limits of the integral; sometimes we also write $a$ and $b$ or $x = a$ and $x = b$.

$$\int_a^b f(x)\,dx = \int_{x=a}^{x=b} f(x)\,dx$$

The area between $x = a$ and $x = a$ is zero, no matter which function you integrate.

$$\int_a^a f(x)\,dx = 0$$

If you first integrate from $a$ to $b$ and then from $b$ to $c$, it's the same as integrating directly from $a$ to $c$.

$$\int_a^b f(x)\,dx + \int_b^c f(x)\,dx = \int_a^c f(x)\,dx$$

If you swap the limits of integration, the sign of the integral changes.

$$\int_b^a f(x)\,dx = -\int_a^b f(x)\,dx$$

Thus, if you integrate from $a$ to $b$ and from $b$ to $a$, you are integrating from $a$ to $a$ and thus the value of the integral is zero.

$$\int_a^b f(x)\,dx + \int_b^a f(x)\,dx = 0$$

The variable of integration can be named as you wish, just like the index of a sum.

$$\int_{x=0}^{1} f(x)\,dx = \int_{u=0}^{1} f(u)\,du$$

# What is an indefinite integral?

An indefinite integral is an integral in which the upper limit is a variable. The area as a function of the upper limit then defines a function, often denoted as $F(x)$.

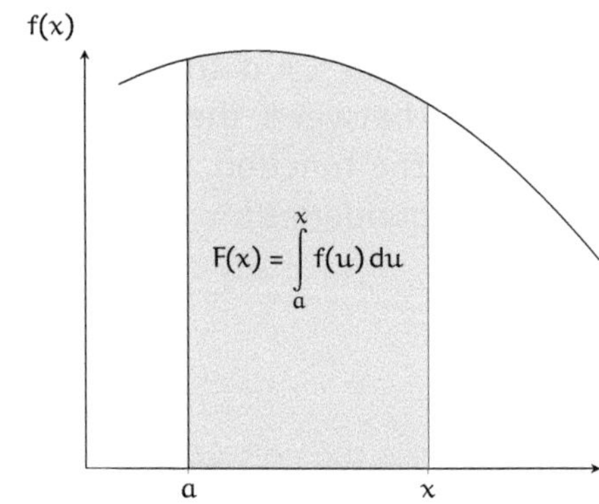



$$F(x) = \int_a^x f(u)\,du$$

$u$: variable of integration
$x$: upper limit

If you denote the variable of the upper limit by $x$, you should use a different symbol for the variable under the integral, otherwise $x$ would appear twice with different meanings.

This is like with sums, where you also use a different index inside than for the upper limit.

$$\sum_1^n a_k, \quad \text{not} \sum_1^n a_n$$

Some people also use the same symbol, in which case the inner variable is what is called a local variable in programming.

Not recommended:

$$F(x) = \int_a^x f(x)\,dx$$

For constant or linear functions, you can determine the area under the function using elementary geometry and thus directly state the integral.

$$\int_0^x c\,dx = cx,$$

since it's the area of a rectangle with side lengths $x$ and $c$.

$$\int_0^x x\,dx = \frac{1}{2}x^2,$$

since it's the area of a right triangle with both legs of length $x$, which is $\frac{1}{2}x \cdot x$.

# What is an antiderivative?

A function whose derivative is f is called an antiderivative of f (since the function is the derivative of it), and it is often denoted by F.

The derivative of $x^2$ is $2x$.
An antiderivative of $2x$ is $x^2$.

$$f(x) = 2x, \quad F(x) = x^2$$
$$F'(x) = f(x)$$

Since the derivative of a constant (a real function that is constant) is zero, you can add any constant to an antiderivative and still have an antiderivative.

The derivative of $x^2 + 17$ is $2x$.
An antiderivative of $2x$ is $x^2 + 17$.

An antiderivative is therefore only determined up to an additive constant, which is often written as "+C".

$$f(x) = 2x \Rightarrow F(x) = x^2 + C$$

This is not a problem, because in concrete calculations you form the difference of the antiderivative at two x-values, and the constant cancels out.

$$F(x_1) - F(x_2) =$$
$$= (x_1^2 + C) - (x_2^2 + C) = x_1^2 - x_2^2$$

The antiderivative can also be written as an integral. The integral sign is a stylized capital letter s for sum, and the x indicates the variable.

$$\int 2x \, dx = x^2 + C$$

Integration is therefore, in a sense, the inverse of differentiation.

# What does the fundamental theorem of calculus state and why is it so useful?

Integrals can be defined as limits of sums, but in general you can calculate an integral faster and more easily using the fundamental theorem of calculus.

The theorem states that if you differentiate the indefinite integral of a function f with respect to the upper limit, you get the original function f back.

$$F(x) = \int_a^x f(u)\,du \Rightarrow F'(x) = f(x)$$

This means that the indefinite integral is simply an antiderivative.

$$\int_a^x f(u)\,du = F(x) + C$$

The definite integral, which you need to calculate area, is thus simply the difference of the antiderivative F at the upper and lower limits.

$$\int_a^b f(x)\,dx = F(b) - F(a)$$

Since antiderivatives of a function differ only by a constant, it does not matter which antiderivative you use.

$$\int_a^b f(x)\,dx =$$
$$= (F(b) + C) - (F(a) + C)$$
$$= F(b) + C - F(a) - C =$$
$$= F(b) - F(a)$$

The difference can also be written with a vertical bar with upper and lower limits, similarly to the integral.

$$\int_a^b f(x)\,dx = F(x)\Big|_a^b = F(b) - F(a)$$

This way, you can easily calculate area without having to form infinite sums.

$$\int_0^1 x^2\,dx = \frac{1}{3}x^3\Big|_0^1 = \frac{1}{3} - \frac{0}{3} = \frac{1}{3}$$

# Why does the fundamental theorem of calculus hold?

We call $F(x)$ the area between the graph of the function $f(x)$ and the x-axis. We are looking for an equation relating the functions $F(x)$ and $f(x)$.

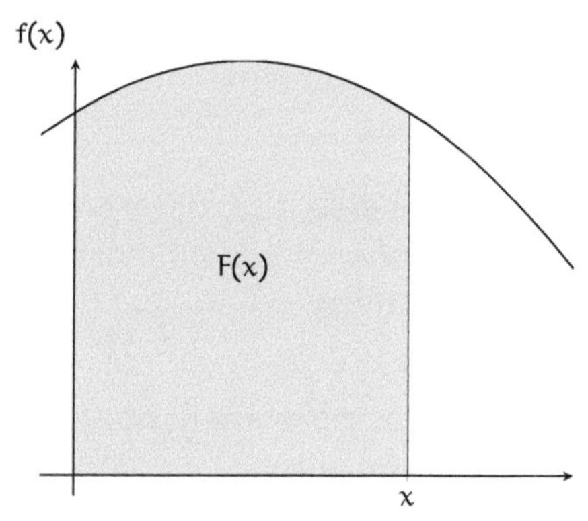


Let us consider how the area changes when $x$ changes by a small amount, from $x$ to $x + h$, where $h$ is supposed to be small. The newly added area is a narrow strip, which is approximately a rectangle with sides $h$ and $f(x)$. The vertical sides $f(x)$ and $f(x + h)$ are approximately, i.e., to first order in $h$, the same. Thus the following doesn't depend on whether you take $f(x)$ or $f(x + h)$ as vertical side.

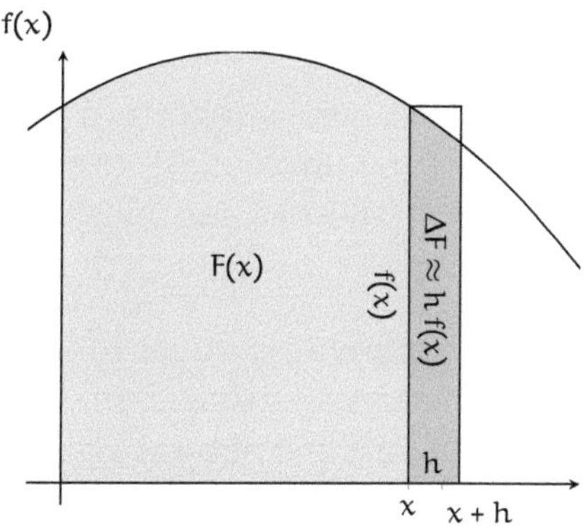


The change in the area is therefore (to first order in $h$) the area of the narrow rectangle, $h\,f(x)$.

$$F(x + h) = F(x) + h\,f(x) + o(h)$$

The change of a function to first order in $h$ is the very definition of the derivative of the function.

$$F(x + h) = F(x) + h\,F'(x) + o(h)$$

From the last two equations, it follows that the derivative of the function that measures the area, $F(x)$, is the function of the graph, $f(x)$.

$$F'(x) = f(x)$$
This is the statement of the fundamental theorem of calculus.

# How do you calculate antiderivatives?

Integrating means finding antiderivatives.

If you reverse the differentiation of functions, you will have their antiderivatives.

$$x^2 \xrightarrow{\text{diff.}} 2x, \quad x^2 \xleftarrow{\text{antideriv.}} 2x$$

$$x^3 \xrightarrow{\text{diff.}} 3x^2, \quad x^3 \xleftarrow{\text{antideriv.}} 3x^2$$

Often you must then multiply or divide by a number to determine the antiderivative of the function at hand.

$$\tfrac{1}{2}x^2 \xrightarrow{\text{diff.}} x, \quad \tfrac{1}{2}x^2 \xleftarrow{\text{antideriv.}} x$$

$$\tfrac{1}{3}x^3 \xrightarrow{\text{diff.}} x^2, \quad \tfrac{1}{3}x^3 \xleftarrow{\text{antideriv.}} x^2$$

In general, you integrate a power function by increasing the exponent by 1 and dividing by this new exponent.

$$\int x^n \, dx = \frac{1}{n+1} x^{n+1} + C$$

If the exponent is equal to −1, this doesn't work. In that case, the antiderivative is the natural logarithm.

$$\int \frac{1}{x} \, dx = \ln(x) + C$$

You can connect the last two integrals if you note that $x^\alpha = \exp(\alpha \ln(x)) \approx 1 + \alpha \ln(x)$ for small $\alpha$ and $x \neq 0$.

If you read a table of derivatives backwards, you get a table of antiderivatives or integrals.

$$\int \cos(x) \, dx = \sin(x) + C$$

If you read differentiation rules backwards, you get integration rules.

E.g. you can find antiderivatives calculating term by term and placing constants before the integral.

$$\int (af(x) + bg(x)) \, dx =$$

$$= a \int f(x) \, dx + b \int g(x) \, dx$$

# How can you calculate areas using antiderivatives?

In the following $F(x)$ will denote the antiderivative of $f(x)$.

If the function is always positive throughout the interval, for calculating the area you simply take the integral, which is the antiderivative at the upper limit minus the antiderivative at the lower limit.

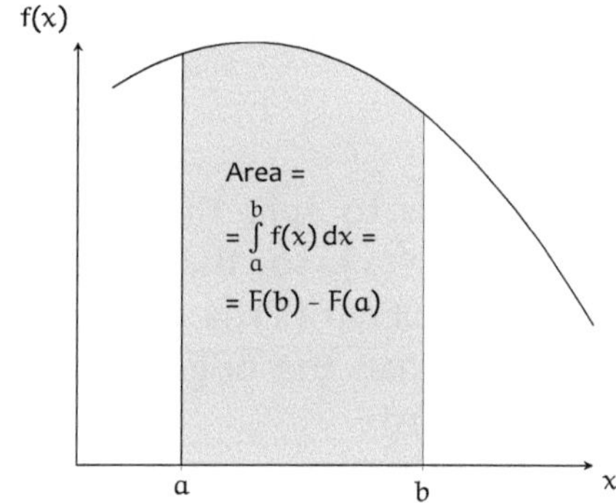


If the function is always negative throughout the interval, you take the negative of the integral (or the absolute value) to obtain a positive area.

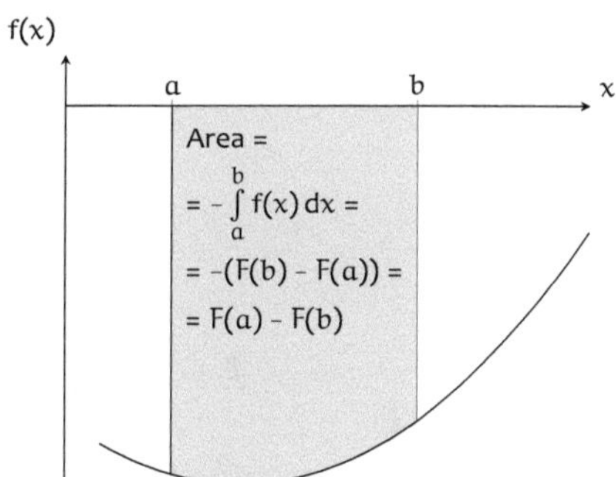


If the sign of the function changes, you integrate from zero to zero (i.e. from one intercept of the x-axis to the next), take the absolute value each time and sum these up.

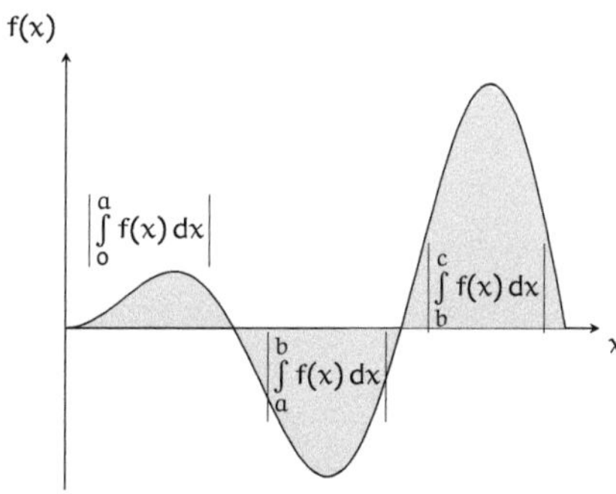

# How do you calculate integrals using integration by parts? (I)

By integrating the product rule, you can obtain the rule of integration by parts. In the following we will omit the constant of integration.

$$(uv)' = u'v + uv'$$

$$uv = \int u'v\,dx + \int vu'\,dx$$

$$\int uv'\,dx = uv - \int vu'\,dx$$

The trick is to split the integrand into $u(x)v'(x)$ so that the new integral $\int v(x)u'(x)\,dx$ is easier to calculate than the original integral $\int u(x)v'(x)\,dx$.

Split the integrand into two factors so that one is easy to integrate and the other is easy to differentiate.

So on the right hand side you choose the first factor $u(x)$, such that it becomes simpler when differentiated.

$$I = \int xe^x dx$$

$$u = x \Rightarrow u' = 1$$

$$v' = e^x \Rightarrow v = e^x$$

$$I = uv - \int vu'\,dx = xe^x - \int e^x dx$$

$$\int xe^x dx = xe^x - e^x$$

You can also remember integration by parts as $\int u\,dv = uv - \int v\,du$.

$$\int u\,dv = uv - \int v\,du$$

$$I = \int xe^x dx$$

$$u = x \Rightarrow du = dx$$

$$dv = e^x dx \Rightarrow v = e^x$$

$$I = uv - \int vu'\,dx = xe^x - \int e^x dx$$

$$\int xe^x dx = xe^x - e^x$$

# How do you calculate integrals using integration by parts? (II)

It can also be helpful to choose the second factor as 1 if the first factor becomes much simpler when differentiated, which is especially the case with the logarithm.

$$I = \int \ln(x)\, dx$$

$$u = \ln(x) \Rightarrow u' = \tfrac{1}{x}$$

$$v' = 1 \Rightarrow v = x$$

$$I = uv - \int vu'\, dx = x\ln(x) - \int x\tfrac{1}{x}\, dx$$

$$= x\ln(x) - \int dx$$

$$= x\ln(x) - x$$

It might be the case that the original integral reappears after rearranging, which is also helpful if you can then solve for it.

$$I = \int \cos^2(x)\, dx = \int \cos(x)\cos(x)\, dx$$

$$u = \cos(x) \Rightarrow u' = -\sin(x)$$

$$v' = \cos(x) \Rightarrow v = \sin(x)$$

$$I = \cos(x)\sin(x) + \int \sin(x)\sin(x)\, dx$$

$$I = \sin(x)\cos(x) + \int (1 - \cos^2(x))\, dx$$

$$I = \sin(x)\cos(x) + \int dx - I$$

$$2I = \sin(x)\cos(x) + x$$

$$I = \frac{1}{2}\left(\sin(x)\cos(x) + x\right)$$

# How do you calculate integrals using substitution? (I)

By integrating the chain rule and swapping the sides, you obtain the substitution rule.

$$(f(g(x))' = f'(g(x))g'(x)$$

$$\int f'(g(x))\, g'(x)\, dx \overset{g(x)=z}{=} \int f'(z)\, dz$$

$$= \quad f(z)$$

The simplest case is when the inner function is a linear function $g(x) = ax + b$.

$$\int f'(ax + b)a\, dx = \int f'(z)\, dz = f(z)$$

$$\Rightarrow \int f'(ax + b)\, dx = \frac{1}{a}f(ax + b)$$

If the inner function is more complicated (e.g. $x^2$), its derivative must appear as a factor in front of the outer function; if necessary, you have to rearrange accordingly.

$$\int x \cos(x^2)\, dx = \int \frac{1}{2} 2x \cos(x^2)\, dx =$$

$$= \frac{1}{2} \int (\cos(x^2))'\, dx = \frac{1}{2} \cos(x^2)$$

You can also replace the term that bothers you (e.g. a complicated term under a root) with a variable such as $z$ and convert all other $x$ into $z$. The conversion between $dx$ and $dz$ is obtained from $z' = \frac{dz}{dx} = g'(x)$. Then you integrate by $z$. At the end, you substitute the expression in $x$ back into $z$.

$$I = \int \frac{x}{\sqrt{1 + x^2}} dx$$

$$1 + x^2 = z \Rightarrow \frac{dz}{dx} = 2x \Rightarrow 2x\, dx = dz$$

$$I = \int \frac{x}{\sqrt{z}} \frac{dz}{2x} =$$

$$= \frac{1}{2} \int \frac{1}{\sqrt{z}} dz = \frac{1}{2} \int z^{-1/2} dz =$$

$$= \frac{1}{2} \frac{1}{1/2} z^{1/2} = z^{1/2} = \sqrt{z} = \sqrt{1 + x^2}$$

You can easily remember the substitution rule if you keep in mind that from $z = g(x)$ and $g'(x) = \frac{dg}{dx}$, the conversion of the differentials follows: $dz = dg = g'(x)dx$.

$$\int f(g(x))\, g'(x)\, dx = \int f(z)\, dz$$

# How do you calculate integrals using substitution? (II)

The substitution rule (see above) can also be read from right to left.

Choose the substitution so that the part that bothers you the most disappears. Trigonometric functions can also be helpful here.

$$\int f(x)\,dx \overset{x=g(t)}{=} \int f'(g(t))\,g'(t)\,dt$$

$$I = \int \sqrt{1 - x^2}\,dx$$

With $x = \sin(t)$, $dx = \cos(t)\,dt$, the root disappears because $\sqrt{1 - \sin^2 t} = \cos t$:

$$I = \int \cos^2(t)\,dt \overset{(*)}{=} \frac{1}{2}(t + \sin(t)\cos(t))$$

$(*)$ integration by parts, see above

$$I = \frac{1}{2}(\arcsin(x) + x\sqrt{1 - x^2})$$

If the numerator is the derivative of the denominator, the integral is simply the logarithm of the denominator. This is a special case of the substitution rule.

$$\int \frac{f'(x)}{f(x)}\,dx = \int \frac{dz}{z} = \ln|z| = \ln|f(x)|$$

$$\int \frac{e^x}{e^x + 1}\,dx = \ln(e^x + 1)$$

$$\int \tan(x)\,dx = -\ln|\cos(x)|$$

For definite integrals it's important that you substitute back, i.e. replace the new variable with the expression in $x$ before plugging in the limits.

$$I = \int_0^1 \sqrt{1 - x^2}\,dx \quad \text{with } x = \sin(t)$$

$$I = \frac{1}{2}(t + \sin(t)\cos(t))$$

$$I = \frac{1}{2}\left(\arcsin(x) + x\sqrt{1 - x^2}\right)\Big|_0^1 = \frac{\pi}{4}$$

Alternatively, you can convert the limits to the new variable. In both ways, you have determined the area of a quarter circle with radius 1 to be $\pi/4$, i.e. the area of the unit circle is $\pi$.

$$I = \int_{x=0}^{x=1} \sqrt{1 - x^2}\,dx \quad \text{with } t = \arcsin(x)$$

$$I = \frac{1}{2}\left(t + \sin(t)\cos(t)\right)\Big|_{t=0}^{t=\pi/2} = \frac{\pi}{4}$$

# How do you calculate the area of a circle by integrating with the radius?

We denote the area of the circle as a function of the radius by $A(x)$.

We are looking for an equation for $A(x)$. To do this, we consider how $A(x)$ changes when $x$ changes a little, from $x$ to $x + h$, where $h$ is small.

The area then changes by the area of an annulus whose inner circumference is $2\pi x$ and whose thickness is $h$.

Thus, the area changes by $2\pi x h$ in first order to $h$.

Therefore, the derivative of $A(x)$ is $2\pi x$.

Thus, you obtain the area from the circumference by integration $A'(x)$ with $x$.

Similarly, you can also determine the volume of a sphere by integrating the surface area with the radius.

radius: $x$
area of circle: $A(x)$

$$A(x + h) = A(x) + \text{area of annulus}$$

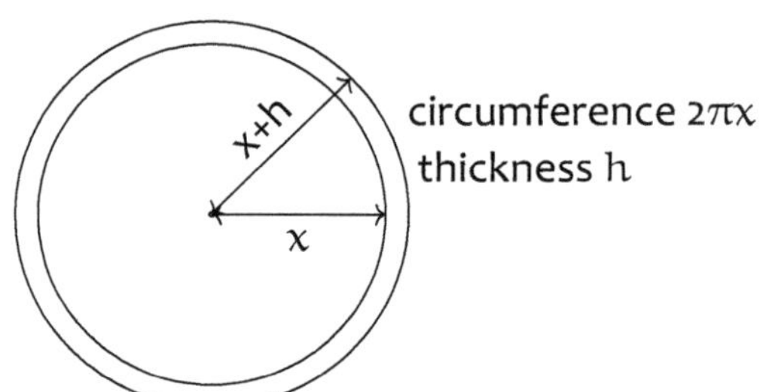


area of the annulus $\approx 2\pi x\, h$

$$A(x + h) = A(x) + \underbrace{2\pi x}_{A'(x)}\, h + o(h^2)$$

$$A'(x) = 2\pi x$$

$$A(r) = \int_{x=0}^{r} 2\pi x\, dx = \pi r^2$$

$$V(r) = \int_{x=0}^{r} S(x)\, dx =$$

$$= \int_{x=0}^{r} 4\pi x^2\, dx = \frac{4}{3}\pi r^3$$

# When is a function integrable?

A function is called integrable if you can assign an area to the region under its graph.

To do this, you can divide the corresponding region on the $x$-axis into intervals and, in each interval, choose a least upper bound and a greatest lower bound, thus approximating the area by an upper sum and a lower sum.

An area exists if, for every sequence of partitions that become arbitrarily fine (measured by the maximum interval width $\Delta$), both the lower sum converges and the upper sum converges, and the two limits are equal.

Note that a function does not have to be differentiable in order to be integrable.

It can be shown that all piecewise continuous functions are integrable. On the right side is a function that is not Riemann integrable.

It's much more difficult to integrate a function, i.e. to explicitly find an antiderivative, than to differentiate a function. There are simple functions without elementary antiderivatives.

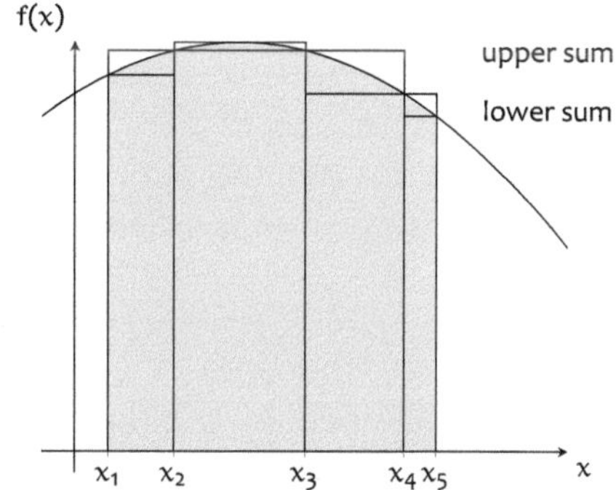



The area ("Riemann integral") exists if

$$\lim_{\Delta \to 0} \text{upper sum} = U \quad \text{exists,}$$

$$\lim_{\Delta \to 0} \text{lower sum} = L \quad \text{exists}$$

and $U = L$ holds.

$f(x) = |x|$ is integrable on $[-1, 1]$, but not differentiable.
Integrability is a weaker requirement.

Counterexample: $f : [0, 1] \to \mathbb{R}$

with $f(x) = \begin{cases} 0 & \text{for} \quad x \in \mathbb{Q} \\ 1 & \text{for} \quad x \in \mathbb{R} \setminus \mathbb{Q} \end{cases}$

is not Riemann integrable.

For example, it can be proven for the functions $\frac{e^x}{x}$ and $e^{-x^2}$ that they do not have an antiderivative that can be expressed in terms of elementary functions. In other words, they are not integrable in closed form.

# When does an improper integral of a function exist?

An integral is called improper if either one or both of the limits of integration are $+\infty$ or $-\infty$, or if the function is not defined at one of the limits of integration and the limit is $+\infty$ or $-\infty$.

In this case, the area is either infinitely extended in the $x$- or the $y$-direction. Nevertheless, the area can be finite. In this case, the improper integral of the function exists.

You calculate the limit by letting one interval bound tend to infinity, or to the value where the function tends to infinity.

Important cases are decreasing power functions $x^\alpha$, $\alpha < 0$. For $\alpha < -1$, the area towards the $x$-axis is finite and the area towards the $y$-axis is infinite.

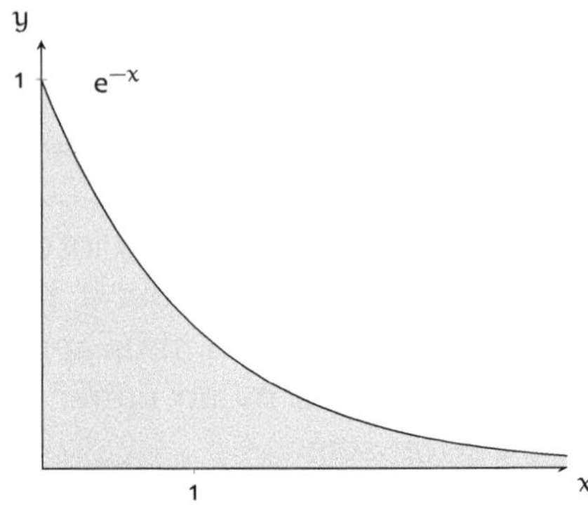



$$\int_0^\infty e^{-x}\, dx = \lim_{a \to \infty} \int_0^a e^{-x}\, dx =$$

$$= \lim_{a \to \infty} \left. -e^{-x}\right|_0^a = \lim_{a \to \infty}(1 - e^{-a}) = 1$$

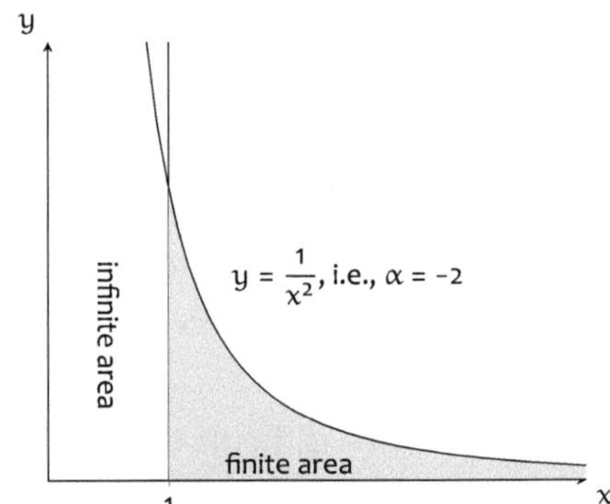



For $\alpha > -1$, it's the other way around.

For $\alpha = -1$, i.e., $f(x) = 1/x$, both areas are infinite.

# Overview: Functions and their antiderivatives

If you read the table of derivatives of simple functions and differentiation
rules backwards, you get the following table of antiderivatives and integration rules.

| $f(x)$ | $\int f(x)\,dx$ |
|---|---|
| $x^n$ | $\dfrac{1}{n+1}x^{n+1},\, n \neq -1$ |
| $\dfrac{1}{x}$ | $\ln|x|$ |
| $e^x$ | $e^x$ |
| $a^x = e^{x\ln(a)}$ | $\dfrac{1}{\ln(a)}a^x$ |
| $\sin(x)$ | $-\cos(x)$ |
| $\cos(x)$ | $\sin(x)$ |
| $\tan(x)$ | $\ln|\cos(x)|$ |
| $af(x) + bg(x)$ | $a\int f(x)\,dx + b\int g(x)\,dx$ |
| $f(x)g'(x)$ | $f(x)g(x) - \int g(x)f'(x)\,dx$ |
| $f(g(x))g'(x)$ | $f(g(x))$ |

# Vector Algebra and Elementary Analytic Geometry

In this chapter, I will explain what vectors are in 2- and 3-dimensional Euclidean space, how to calculate with them (vector algebra), and how you can use them to set up equations for lines and planes in 2 and 3 dimensions (analytic geometry).

Important concepts are linear independence and linear dependence of a set of vectors, and the scalar product, which assigns a number to two vectors.

The next chapter will then deal with general vector spaces, linear mappings, matrices, determinants, the cross product, and how you can solve systems of linear equations. With this, you will be able to determine intersection points of affine subspaces (such as lines and planes).

# What is linear algebra and analytic geometry?

A central subject of linear algebra is vector spaces and their structure, as well as linear mappings.

$$\mathbf{v} \in \mathbb{R}^2$$
$$\uparrow \qquad \uparrow$$


Vectors are often written in bold, $\mathbf{v}$, which is also what I do here. In handwriting, it is common to draw an arrow above the symbol, i.e. $\vec{v}$.

Some people underline them; sometimes vectors are not marked at all. It's best to always check the domain of definition of the objects.

You can add vectors and multiply them by numbers (called scalars) which we call scalar multiplication.

$$\begin{pmatrix} u_1 \\ u_2 \end{pmatrix} + \begin{pmatrix} v_1 \\ v_2 \end{pmatrix} = \begin{pmatrix} u_1 + v_1 \\ u_2 + v_2 \end{pmatrix}$$

$$s \begin{pmatrix} v_1 \\ v_2 \end{pmatrix} = \begin{pmatrix} sv_1 \\ sv_2 \end{pmatrix}$$

Structure-preserving mappings between vector spaces, so-called linear mappings, and their representations play an important role in linear algebra.

$$f : \mathbb{R}^2 \to \mathbb{R}^2$$

$$\begin{pmatrix} v_1 \\ v_2 \end{pmatrix} \mapsto \begin{pmatrix} a_{11}v_1 + a_{12}v_2 \\ a_{21}v_1 + a_{22}v_2 \end{pmatrix}, a_{ij} \in \mathbb{R}$$

Vectors can "live" in Euclidean space, but, for example, functions can also be regarded as vectors; for instance, the set of all quadratic functions from $\mathbb{R}$ to $\mathbb{R}$ forms a 3-dimensional vector space.

The sum of two quadratic functions is again a quadratic function. Likewise, a quadratic function remains a quadratic function when you multiply it by a number.

You can multiply two vectors with each other, even in two different ways.

Scalar multiplication yields a number; vector multiplication yields a vector.

In analytic geometry, geometric objects such as lines, planes, circles and spheres are represented using vectors.

Line $g$ through $A$ with direction $\mathbf{u}$:
$$g: \mathbf{x} = \mathbf{x}_A + s\mathbf{u}, \ s \in \mathbb{R}$$

Sphere with center $M$ and radius $r$:
$$K: \left| \mathbf{x} - \mathbf{x}_M \right| = r$$

# What is a vector intuitively? (I)

A position vector is an arrow (directed segment) from the origin to a point in the plane (or in space). For $\overrightarrow{OA}$, we also write $\mathbf{x}_A$ or simply $\mathbf{a}$.

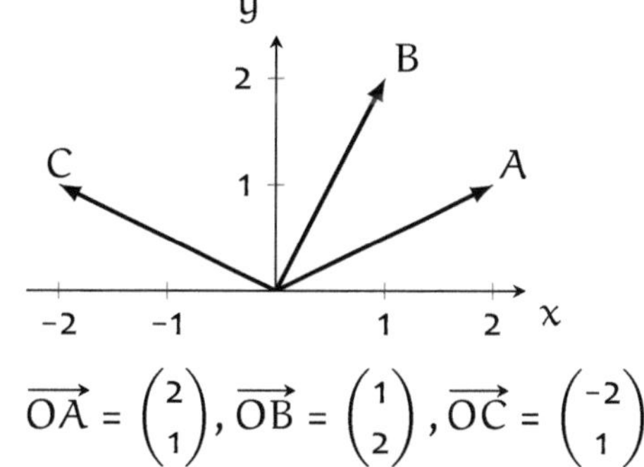

The components of the position vector are the coordinates of the endpoint.

$$\overrightarrow{OA} = \begin{pmatrix} 2 \\ 1 \end{pmatrix}, \overrightarrow{OB} = \begin{pmatrix} 1 \\ 2 \end{pmatrix}, \overrightarrow{OC} = \begin{pmatrix} -2 \\ 1 \end{pmatrix}$$

A vector is determined by its length and direction; it can also start at a point other than the origin. In other words, it can be shifted parallel anywhere without the vector being altered.

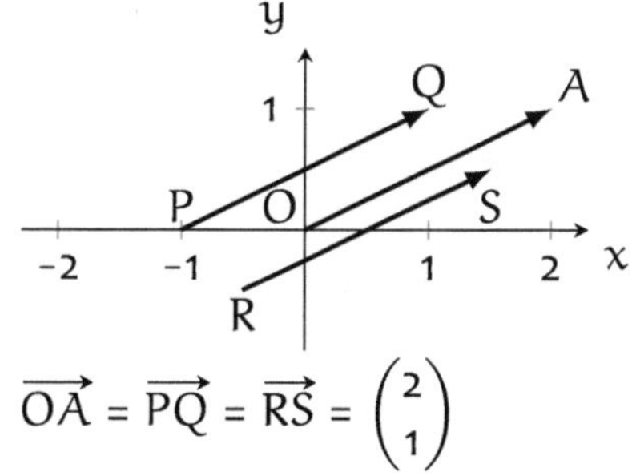

$$\overrightarrow{OA} = \overrightarrow{PQ} = \overrightarrow{RS} = \begin{pmatrix} 2 \\ 1 \end{pmatrix}$$

Its components are the coordinates of the endpoint minus the coordinates of the starting point.

$$\overrightarrow{AB} = \begin{pmatrix} x_B - x_A \\ y_B - y_A \end{pmatrix}$$

In three dimensions, this works in exactly the same way.

$$\overrightarrow{AB} = \begin{pmatrix} x_B - x_A \\ y_B - y_A \\ z_B - z_A \end{pmatrix}$$

We write vectors as column vectors. If we write a column vector in a single line to save space, we add the transpose symbol ($\top$). It turns rows into columns and vice versa.

$$(a_1, a_2, a_3)^\top = \begin{pmatrix} a_1 \\ a_2 \\ a_3 \end{pmatrix}$$

$$\begin{pmatrix} a_1 \\ a_2 \\ a_3 \end{pmatrix}^\top = (a_1, a_2, a_3)$$

# What is a vector intuitively? (II)

The components of a vector are the coordinates of the endpoint minus the coordinates of the starting point.

$$\overrightarrow{AB} = \begin{pmatrix} x_B - x_A \\ y_B - y_A \end{pmatrix}$$

If the starting point is $O = (0, 0)$ or $O = (0, 0, 0)$ we get the above definition of a position vector.

$$\overrightarrow{OP} = \begin{pmatrix} x_P - 0 \\ y_P - 0 \end{pmatrix} = \begin{pmatrix} x_P \\ y_P \end{pmatrix}$$

The set of all vectors in the plane is called the 2-dimensional vector space $\mathbb{R}^2$.

$$\mathbf{v} \in \mathbb{R}^2$$

The 2 in the exponent indicates that you can choose two real numbers, the two components $v_1$ and $v_2$, independently of each other.

$$\begin{pmatrix} v_1 \\ v_2 \end{pmatrix} \in \mathbb{R}^2$$

In 3-dimensional space this works exactly analogously to the 2-dimensional plane.

$$\begin{pmatrix} v_1 \\ v_2 \\ v_3 \end{pmatrix} \in \mathbb{R}^3$$

The length of a vector is often denoted by the same symbol, but without boldface, or with single or double bars.

$|\mathbf{a}| = ||\mathbf{a}|| = a$: length of the vector $\mathbf{a}$.

The length of a vector is also called the norm of the vector.
If it has length 1, it is called normalized.

The length is calculated using the Pythagorean theorem, applied twice to result in 3 dimensions.

$$|\mathbf{a}| = \sqrt{a_1^2 + a_2^2} \qquad \text{in 2 dimensions}$$
$$|\mathbf{a}| = \sqrt{a_1^2 + a_2^2 + a_3^2} \quad \text{in 3 dimensions}$$

In 1 dimension, a vector has only one component, so the length is $\sqrt{a^2} = |a|$, i.e. the absolute value of its only component.

# How do you multiply vectors by a real number?

If you multiply a vector by a real number $r > 0$, the direction of the vector remains the same and the length is multiplied by $r$.

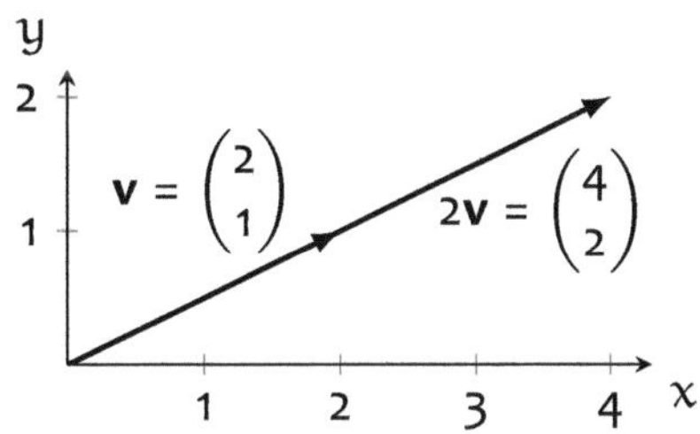

You multiply each component of the vector by the number.

$$rv = r\begin{pmatrix} v_1 \\ v_2 \end{pmatrix} = \begin{pmatrix} rv_1 \\ rv_2 \end{pmatrix} = \begin{pmatrix} r \cdot 2 \\ r \cdot 1 \end{pmatrix} = \begin{pmatrix} 4 \\ 2 \end{pmatrix}$$

If the real number is less than one, the resulting vector is shorter.

$$\tfrac{1}{2}v = \begin{pmatrix} 1 \\ \tfrac{1}{2} \end{pmatrix}$$

If the real number is zero, the resulting vector is the zero vector.

$$0v = \begin{pmatrix} 0 \\ 0 \end{pmatrix} = 0$$

When you multiply a vector $v$ by $-1$, you get $-v$, i.e. the vector of the same length in the opposite direction.

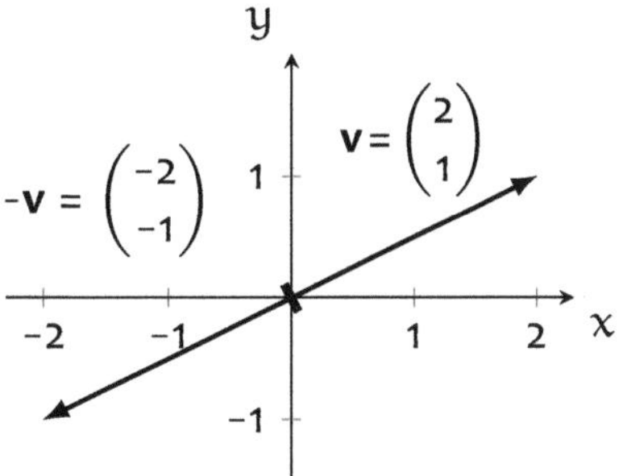

If you multiply a vector by a real number $r < 0$, the direction of the vector is reversed and its length is multiplied by $|r|$.

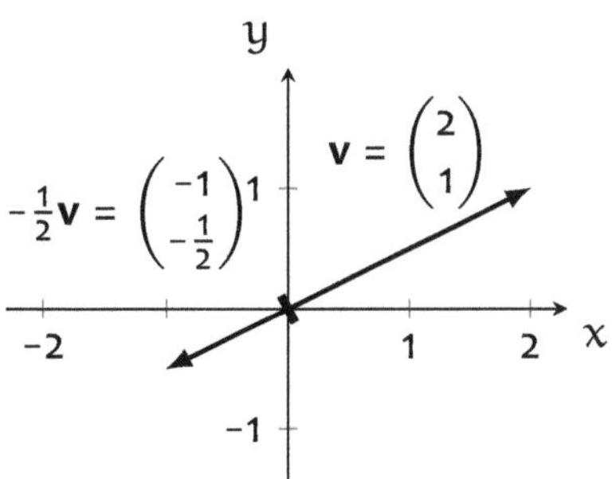

# How do you add and subtract vectors?

Two vectors, where the endpoint of the first vector is the same as the starting point of the second vector, have as their sum the vector that goes from the starting point of the first vector to the endpoint of the second vector.

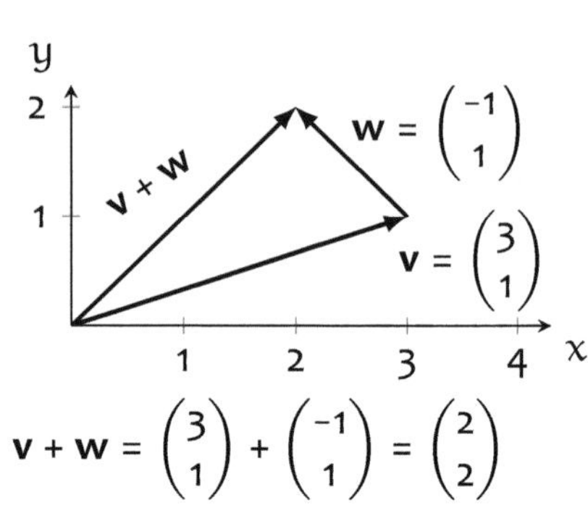


$$v + w = \begin{pmatrix} 3 \\ 1 \end{pmatrix} + \begin{pmatrix} -1 \\ 1 \end{pmatrix} = \begin{pmatrix} 2 \\ 2 \end{pmatrix}$$

When you add any two vectors, you shift one vector so that its starting point coincides with the endpoint of the other vector. The sum is then again the vector from the starting point of the first vector to the endpoint of the second.

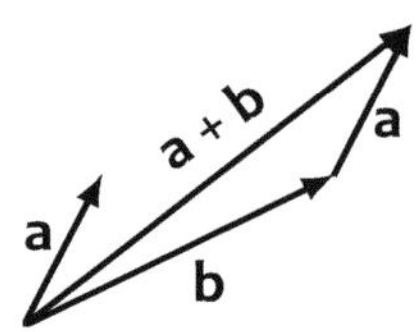


By considering a parallelogram, you can see that the order in which you add vectors doesn't matter: the endpoint is the same whether you go along vector **a** first and then vector **b**, or the other way around.

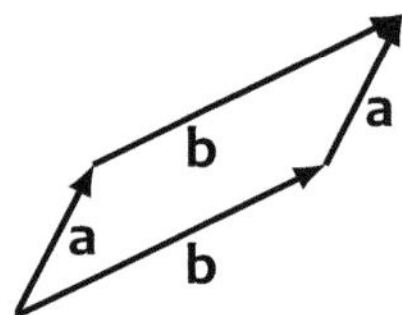


The difference vector **w** − **v** goes from the endpoint of **v** to the endpoint of **w**.

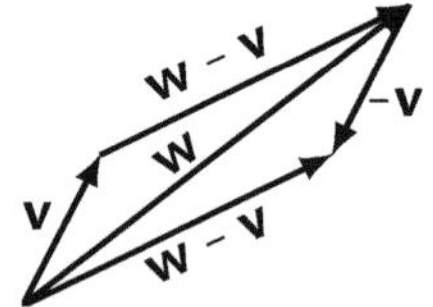


Equivalently, we can add the negative of **v** to **w**: **w** − **v** = **w** + (−**v**).

# What is a linear combination of vectors?

A linear combination of vectors is a (finite) sum of multiples of these vectors, where, of course, negative multiples are also allowed, so differences as well.

$2\mathbf{a} + 3\mathbf{b}$ is a linear combination of $\mathbf{a}$ and $\mathbf{b}$.

$5.6\mathbf{a} - 17.3\mathbf{b} - \mathbf{c}$ is a linear combination of $\mathbf{a}$, $\mathbf{b}$ and $\mathbf{c}$.

The set of all linear combinations of a set of vectors is also called the linear space spanned by them.

$\{r_1\mathbf{a}_1 + r_2\mathbf{a}_2 + \ldots + r_n\mathbf{a}_n, r_i \in \mathbb{R}\}$ is the space spanned by $\mathbf{a}_1, \mathbf{a}_2, \ldots, \mathbf{a}_n$. This is also called the linear span, or just span, or linear hull.

For two vectors the span can look very different depending on the vectors spanning it.

The two vectors $(0,0,1)^\top$ and $(0,1,0)^\top$ span the 2-dimensional plane $\left\{(0,r,s)^\top, r,s \in \mathbb{R}\right\}$.

The two vectors $(0,0,1)^\top$ and $(0,0,-1)^\top$ span the 1-dimensional line $\left\{(0,0,r)^\top, r \in \mathbb{R}\right\}$.

The two vectors $(0,0,0)^\top$ and $(0,0,0)^\top$ span the 0-dimensional vector space $\left\{(0,0,0)^\top\right\}$.

This holds in general: The dimension of the span of $n$ vectors can be 0-, 1-, ..., $n$-dimensional, depending on how the $n$ vectors are positioned relative to each other.

The $n$ vectors

$$\begin{pmatrix} 1 \\ 0 \\ \vdots \\ 0 \end{pmatrix}, \begin{pmatrix} 0 \\ 1 \\ \vdots \\ 0 \end{pmatrix}, \ldots, \begin{pmatrix} 0 \\ 0 \\ \vdots \\ 1 \end{pmatrix}$$

generate the $n$-dimensional space.

The dimension of the span of $n$ vectors is exactly $n$ if and only if the set of $n$ vectors is linearly independent.

The set of these vectors is linearly independent, they form a basis of $\mathbb{R}^n$.

# When is a set of vectors linearly dependent?

A set of vectors is linearly dependent if you can form a nontrivial linear combination that yields the zero vector.

$$r_1 \mathbf{a_1} + r_2 \mathbf{a_2} + \ldots + r_n \mathbf{a_n} = \mathbf{0}$$
and not all $r_i = 0$

Nontrivial means that at least one of the coefficients is not zero.

Two identical vectors are linearly dependent.

$$1 \cdot \mathbf{a} + (-1) \cdot \mathbf{a} = \mathbf{0}$$
$\Rightarrow$ **a** and **a** are linearly dependent.

Here and in the following, we write a multiplication dot for clarity.

Two parallel or antiparallel vectors are linearly dependent.

$$r \cdot \mathbf{a} + (-1) \cdot (r\mathbf{a}) = \mathbf{0}$$
$\Rightarrow$ **a** and $r$**a** are linearly dependent.

$r$ and $-1$ are the coefficients.

A set of vectors that contains the zero vector **0** is always linearly dependent.

$$0 \cdot \mathbf{a} + 0 \cdot \mathbf{b} + 1 \cdot \mathbf{0} = \mathbf{0}$$
$\Rightarrow$ **a**, **b** and **0** are linearly dependent.

Three vectors in the plane are always linearly dependent.

This results from the fact that two vectors **u** and **v**, if they are not already linearly dependent, span the entire plane, i.e. every point in the plane **x** can be written as a linear combination of **u** and **v**: $\mathbf{x} = r\mathbf{u} + s\mathbf{v}$.
So $r\mathbf{u} + s\mathbf{v} - 1\mathbf{x} = 0$.

$n$ vectors are linearly independent if the dimension of the space they span is $n$.

The two vectors $\begin{pmatrix} 0 \\ 1 \end{pmatrix}, \begin{pmatrix} 1 \\ 1 \end{pmatrix}$ span the 2-dimensional space

$$\left\{ \begin{pmatrix} r \\ s \end{pmatrix}, r, s \in \mathbb{R} \right\} \text{ and are}$$

therefore linearly independent.

If the dimension of the space they span is less than $n$, the vectors are linearly dependent.

The two vectors $\begin{pmatrix} 0 \\ 1 \end{pmatrix}, \begin{pmatrix} 0 \\ 2 \end{pmatrix}$ span the 1-dimensional space

$$\left\{ \begin{pmatrix} 0 \\ r \end{pmatrix}, r \in \mathbb{R} \right\} \text{ and are therefore}$$

linearly dependent.

# When is a set of vectors linearly independent and how do you check that?

A set of vectors is linearly independent if you can produce the zero vector only by a trivial linear combination.

If the following holds:

$$r_1 \mathbf{a}_1 + r_2 \mathbf{a}_2 + \ldots + r_n \mathbf{a}_n = \mathbf{0}$$
$$\Rightarrow r_i = 0 \text{ for all } i,$$

then the vectors $\mathbf{a}_1, \mathbf{a}_2, \ldots, \mathbf{a}_n$ are linearly independent.

Intuitively, you can say that the mapping from the coefficients to the linear combination is then injective. For a given vector as a linear combination, there is a unique tuple of coefficients; for the zero vector as a linear combination, all coefficients are zero.

To check whether a set of vectors is linearly independent, you consider a general linear combination of them and equal this combination to zero. If you can show that this implies that all coefficients of the linear combination must be zero, you have shown that the vectors are linearly independent.

Is $\{(2, 3)^\top, (3, 5)^\top\}$ lin. independent?

From $r \begin{pmatrix} 2 \\ 3 \end{pmatrix} + s \begin{pmatrix} 3 \\ 5 \end{pmatrix} = \begin{pmatrix} 0 \\ 0 \end{pmatrix}$ follows

$$2r + 3s = 0$$
$$3r + 5s = 0.$$

Solving the system gives $r = s = 0$.

$\Rightarrow$ The vectors $\begin{pmatrix} 2 \\ 3 \end{pmatrix}$ and $\begin{pmatrix} 3 \\ 5 \end{pmatrix}$

are linearly independent.

If there is a nontrivial solution for the coefficients, the vectors are linearly dependent.

Is $\{(2, 3)^\top, (4, 6)^\top\}$ lin. independent?

From $r \begin{pmatrix} 2 \\ 3 \end{pmatrix} + s \begin{pmatrix} 4 \\ 6 \end{pmatrix} = \begin{pmatrix} 0 \\ 0 \end{pmatrix}$ follows

$$2r + 4s = 0$$
$$3r + 6s = 0.$$

A solution is e.g. $r = 2, s = -1$.

$\Rightarrow$ The vectors $\begin{pmatrix} 2 \\ 3 \end{pmatrix}$ and $\begin{pmatrix} 4 \\ 6 \end{pmatrix}$

are linearly dependent.

# What is the dot product?

The dot product, also known as scalar product, assigns a number to two vectors.

$$\mathbf{a}, \mathbf{b} \in \mathbb{R}^2, \quad \mathbf{a} \cdot \mathbf{b} \in \mathbb{R}$$

A number is also called a scalar.

You calculate it by multiplying the respective components and adding these products.

$$\mathbf{a} \cdot \mathbf{b} = a_1 b_1 + a_2 b_2$$

$$\begin{pmatrix} 1 \\ 2 \end{pmatrix} \cdot \begin{pmatrix} 2 \\ 3 \end{pmatrix} = 1 \cdot 2 + 2 \cdot 3 = 8$$

If the dot product is zero, without either vector being the zero vector, it indicates that the two vectors are perpendicular to each other.

$$\mathbf{a} \cdot \mathbf{b} = 0 \Rightarrow \mathbf{a} \perp \mathbf{b}$$

The two vectors $\mathbf{a} = (1, 2)^\top$ and $\mathbf{b} = (-2, 1)^\top$ are perpendicular to each other as $\mathbf{a} \cdot \mathbf{b} = 0$. You can see this geometrically if you draw the parallels to the $y$-axis and notice that the two triangles are congruent to each other.

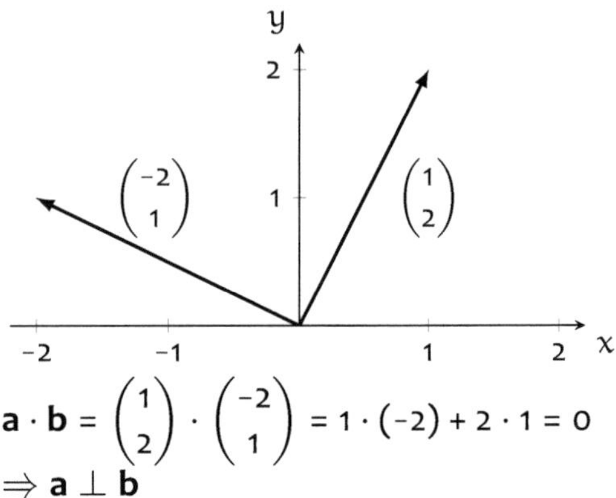

$$\mathbf{a} \cdot \mathbf{b} = \begin{pmatrix} 1 \\ 2 \end{pmatrix} \cdot \begin{pmatrix} -2 \\ 1 \end{pmatrix} = 1 \cdot (-2) + 2 \cdot 1 = 0$$
$$\Rightarrow \mathbf{a} \perp \mathbf{b}$$

In three dimensions it works in exactly the same way.

$$\mathbf{a}, \mathbf{b} \in \mathbb{R}^3, \quad \mathbf{a} \cdot \mathbf{b} \in \mathbb{R}$$
$$\mathbf{a} \cdot \mathbf{b} = a_1 b_1 + a_2 b_2 + a_3 b_3$$

The three vectors $\mathbf{a} = (1, 1, 2)^\top$, $\mathbf{b} = (1, 1, -1)^\top$ and $\mathbf{c} = (-1, 1, 0)^\top$ are pairwise perpendicular to each other.

For example,
$$\mathbf{a} \cdot \mathbf{b} = 1 \cdot 1 + 1 \cdot 1 + 2 \cdot (-1) = 0$$
$$\Rightarrow \mathbf{a} \perp \mathbf{b}$$

You can't solve the scalar product equation $\mathbf{a} \cdot \mathbf{b} = c$ for the vector $\mathbf{a}$, because any vector $\mathbf{b}_\perp$ that is perpendicular to $\mathbf{b}$ will have a scalar product of zero with $\mathbf{b}$. Thus, for any solution $\mathbf{a}$, the vector $\mathbf{a} + \mathbf{b}_\perp$ is also a solution.

The set of endpoints of all vectors $\mathbf{x}$ with $\mathbf{x} \cdot \mathbf{b} = c$ is, in two dimensions, a line perpendicular to $\mathbf{b}$ and, in three dimensions, a plane perpendicular to $\mathbf{b}$. Therefore, there is no unique solution.

# How do you determine the angle between two vectors?

The scalar product is equal to the length of the first vector times the length of the second vector times the cosine of the angle formed by them.

Thus, you can easily determine the angle between two vectors using the scalar product.

To do so, you calculate the length as the square root of the scalar product of the vector with itself.

With this, you can solve the equation of the scalar product above for the cosine of the angle.

If the scalar product is zero, then the cosine is zero and the angle is $\pm 90°$.

You can calculate e.g. the angle between the diagonal of a square and one of its sides.

Similarly, you can calculate the angle between the space diagonal of a cube and one of its edges.

You can understand this by comparing the law of cosines with $(\mathbf{a} - \mathbf{b})^2$:

$$c^2 = a^2 + b^2 - 2ab \cos \varphi$$
$$c^2 = (\mathbf{a} - \mathbf{b})^2 = \underbrace{\mathbf{a}^2}_{a^2} + \underbrace{\mathbf{b}^2}_{b^2} - 2\,\mathbf{a}\cdot\mathbf{b}$$

$$\Rightarrow \mathbf{a} \cdot \mathbf{b} = ab \cos \varphi, \quad \varphi = \angle(\mathbf{a}, \mathbf{b})$$

$$\Rightarrow \cos \varphi = \frac{\mathbf{a} \cdot \mathbf{b}}{ab}$$

$$|\mathbf{a}| = a = \sqrt{\mathbf{a} \cdot \mathbf{a}} = \sqrt{a_1^2 + a_2^2}$$
$$\text{or } \sqrt{a_1^2 + a_2^2 + a_3^2} \text{ in 3D}$$

$$\cos \varphi = \frac{\mathbf{a} \cdot \mathbf{b}}{ab} = \frac{a_1 b_1 + a_2 b_2}{\sqrt{a_1^2 + a_2^2}\sqrt{b_1^2 + b_2^2}}$$

$$\cos \varphi = 0 \Rightarrow \varphi = \pm 90°$$

$$\begin{pmatrix} 1 \\ 1 \end{pmatrix} \cdot \begin{pmatrix} 1 \\ 0 \end{pmatrix} = 1 = \left|\begin{pmatrix} 1 \\ 1 \end{pmatrix}\right|\left|\begin{pmatrix} 1 \\ 0 \end{pmatrix}\right| \cos \varphi$$
$$\Rightarrow 1 = \sqrt{2} \cos \varphi \Rightarrow \cos \varphi = \frac{\sqrt{2}}{2}$$
$$\Rightarrow \varphi = 45°$$

$$\begin{pmatrix} 1 \\ 1 \\ 1 \end{pmatrix} \cdot \begin{pmatrix} 1 \\ 0 \\ 0 \end{pmatrix} = 1 \cdot 1 + 1 \cdot 0 + 1 \cdot 0 = 1$$

$$= \left|\begin{pmatrix} 1 \\ 1 \\ 1 \end{pmatrix}\right|\left|\begin{pmatrix} 1 \\ 0 \\ 0 \end{pmatrix}\right| \cos \varphi$$

$$\Rightarrow 1 = \sqrt{3} \cos \varphi \Rightarrow \cos \varphi = \frac{\sqrt{3}}{3}$$
$$\Rightarrow \varphi = \arccos \frac{\sqrt{3}}{3} \approx 54.7°$$

# What does the dot product have to do with projection?

If the vector $\mathbf{e}$, with which you take the dot product $\mathbf{e} \cdot \mathbf{a}$, is a unit vector, then the dot product is the projection of $\mathbf{a}$ onto the direction of this vector.

$$\mathbf{e} \cdot \mathbf{a} = |\mathbf{e}|\,|\mathbf{a}| \cos \varphi$$
$$|\mathbf{e}| = 1 \Rightarrow \mathbf{e} \cdot \mathbf{a} = |\mathbf{a}| \cos \varphi = a_{\|\mathbf{e}}$$

Thus, $\mathbf{e} \cdot \mathbf{a} = a_{\|\mathbf{e}}$, where $a_{\|\mathbf{e}}$ is the length of the projection of the vector $\mathbf{a}$ onto the direction $\mathbf{e}$.

If you project onto three mutually perpendicular unit vectors, you get three projections, and by multiplying each of them with $\mathbf{e}_i$ you get the corresponding components of the vector along the unit vectors, and by summing them up you recover the original vector.

$$\mathbf{e}_1 \cdot \mathbf{a} = a_{\|\mathbf{e}_1}$$
$$\mathbf{e}_2 \cdot \mathbf{a} = a_{\|\mathbf{e}_2}$$
$$\mathbf{e}_3 \cdot \mathbf{a} = a_{\|\mathbf{e}_3}$$

$$\mathbf{a} = \sum_{i=1}^{3} \mathbf{e}_i\, a_{\|\mathbf{e}_i}$$

For the following, you will need matrix multiplication.

This can be elegantly written using the matrix notation of the dot product as shown on the right, where $\mathbf{1}$ denotes the 3-dimensional identity matrix.

$$\mathbf{a} = \sum_{i=1}^{3} \mathbf{e}_i\, a_{\|\mathbf{e}_i}$$
$$= \sum_{i=1}^{3} \mathbf{e}_i (\mathbf{e}_i \cdot \mathbf{a})$$
$$= \sum_{i=1}^{3} \mathbf{e}_i \mathbf{e}_i^{\top} \mathbf{a}, \text{ so}$$
$$\mathbf{1} = \sum_{i=1}^{3} \mathbf{e}_i \mathbf{e}_i^{\top}.$$

If the $\mathbf{e}_i$ are the canonical basis vectors, you get the simple equation on the right.

$$\begin{pmatrix} 1 & 0 & 0 \\ 0 & 1 & 0 \\ 0 & 0 & 1 \end{pmatrix} = \begin{pmatrix} 1 & 0 & 0 \\ 0 & 0 & 0 \\ 0 & 0 & 0 \end{pmatrix} + \begin{pmatrix} 0 & 0 & 0 \\ 0 & 1 & 0 \\ 0 & 0 & 0 \end{pmatrix} + \begin{pmatrix} 0 & 0 & 0 \\ 0 & 0 & 0 \\ 0 & 0 & 1 \end{pmatrix}$$

This representation of the identity $\mathbf{1}$ using a basis (on the far right in the so-called bra-ket notation, which we will not discuss) can be generalized in many ways.

$$\mathbf{1} = \sum_{i} \mathbf{e}_i \mathbf{e}_i^{\top} \qquad \mathbf{1} = \sum_{i} |i\rangle\langle i|$$

If the vector space is a function space, and the dot product is the integral over the product of two functions, then, with the appropriate complete orthonormal system, you can obtain the Fourier decomposition.

# How can you represent a line in parametric form?

You can specify a line by a point $A$ and the direction.

The position vector to any point $P$ on the line minus the position vector to the given point $A$ is a multiple of the direction vector: $\mathbf{x}_P - \mathbf{x}_A = r\mathbf{v}$.

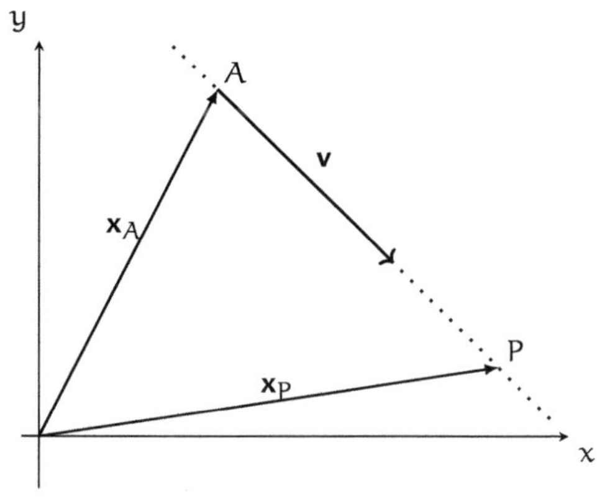


From this, you directly obtain the parametric equation of the line.

$$\mathbf{x}_P = \mathbf{x}_A + r\mathbf{v}, \quad r \in \mathbb{R}$$

For any value of the parameter $r \in \mathbb{R}$ you get a point on the line. As $r$ runs through all real numbers, the corresponding point traces out the entire line.

$r = 0$ gives the position vector $\mathbf{x}_A$, that is, the point $A$,

$r = \frac{1}{2}$ gives $\mathbf{x}_A + \frac{1}{2}\mathbf{v}$

$r = 1$ gives $\mathbf{x}_A + \mathbf{v}$

$r = -1$ gives $\mathbf{x}_A - \mathbf{v}$

If you want to write the parametric form of the line through the two points $A$ and $B$, you can simply choose the difference of the two position vectors as the direction vector.

direction vector:
$$\mathbf{v} = \mathbf{x}_B - \mathbf{x}_A$$

line equation:
$$\mathbf{x}_P = \mathbf{x}_A + r(\mathbf{x}_B - \mathbf{x}_A), \quad r \in \mathbb{R}$$

The parameter value $r = 0$ gives the point $A$, $r = 1$ gives the point $B$.

$$r = 0 \Rightarrow \mathbf{x}_P = \mathbf{x}_A$$
$$r = 1 \Rightarrow \mathbf{x}_P = \mathbf{x}_A + (\mathbf{x}_B - \mathbf{x}_A) = \mathbf{x}_B$$

This applies exactly the same in two and three dimensions, or in any dimension.

In one dimension, it is just a translation and scaling of the number line.

You can of course also write the vectors in components, as on the right hand side for the line above.

$$\begin{pmatrix} x \\ y \end{pmatrix} = \begin{pmatrix} 1 \\ 2 \end{pmatrix} + r \begin{pmatrix} 1 \\ -1 \end{pmatrix}, \quad r \in \mathbb{R}$$

$$\begin{aligned} x &= 1 + r \\ y &= 2 - r \end{aligned} \quad r \in \mathbb{R}$$

# How can you represent a plane in parametric form?

The parametric equation of a plane in 3 dimensions is completely analogous to that of a line; you now just need two parameters.

A plane can be specified by a point and two direction vectors.

The position vector to any point P on the plane minus the position vector to the given point A is a linear combination of the two direction vectors.

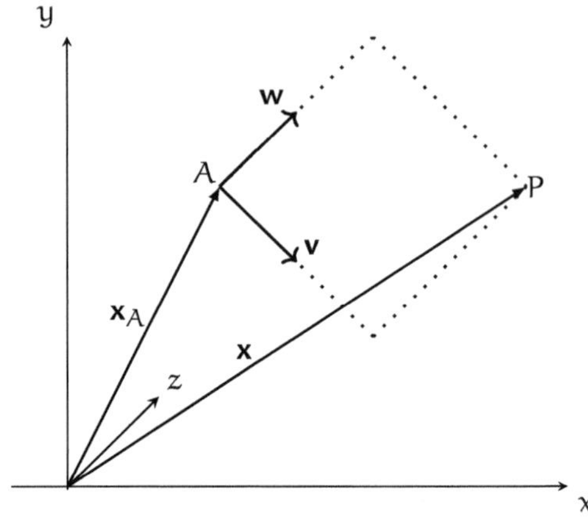


$$x_P - x_A = rv + sw, \qquad r, s \in \mathbb{R}$$
$$x_P = x_A + rv + sw, \qquad r, s \in \mathbb{R}$$

From this you directly obtain the parametric equation of the plane.

$r$ and $s$ are the two parameters; for each pair $(r, s) \in \mathbb{R}^2$ you get a point. As $r$ and $s$ covers through all real numbers, the corresponding point runs through the entire plane.

If you want to specify the parametric form of the plane through three points A, B, C, you can simply choose as direction vectors the two differences of the position vectors of B and C with the position vector of A.

$1^{\text{st}}$ direction vector: $v = x_B - x_A$
$2^{\text{nd}}$ direction vector: $w = x_C - x_A$
$$x_P = x_A + r(x_B - x_A) + s(x_C - x_A)$$
$$r, s \in \mathbb{R}$$

The parameters $r = 0, s = 0$ yield A,
the parameters $r = 1, s = 0$ yield B
and the parameters $r = 0, s = 1$ yield C.

This holds in three dimensions. In two dimensions, the parametric equation of the plane is just a coordinate transformation.

(and also in higher dimensions)

In one dimension, there are no two linearly independent direction vectors and thus no planes.

# How can you represent a line in Hesse normal form? (I)

If you specify any vector **n** in the plane and consider where the tips of all vectors **x** lie, that form a scalar product equal to zero with it, then you get the line through the origin perpendicular to **n**.

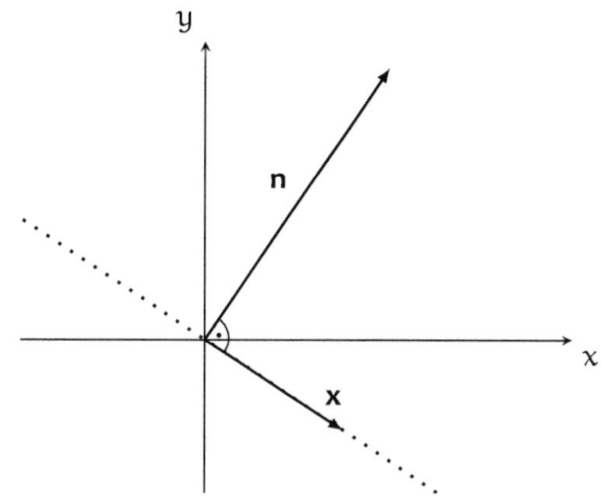

$\mathbf{x} \cdot \mathbf{n} = 0$: equation of the line perpendicular to **n**.

**n**: normal vector of the line. "normal" means perpendicular, i.e. an angle of $90°$.

Therefore, in two dimensions, $\mathbf{x} \cdot \mathbf{n} = 0$ is the equation of the line through the origin perpendicular to **n**.

If the line is still supposed to be perpendicular to **n**, but should pass through the point $A$, then $\overrightarrow{AP} = \mathbf{x} - \mathbf{x}_A$ must be perpendicular to **n**.

Thus all points on the line have a constant scalar product with the normal vector: $\mathbf{x} \cdot \mathbf{n} = \mathbf{x}_A \cdot \mathbf{n} = \text{const.}$ This is consistent with the fact that the scalar product $\mathbf{x} \cdot \mathbf{n}$ is the product of the projection of **x** onto **n** and the length of **n**.

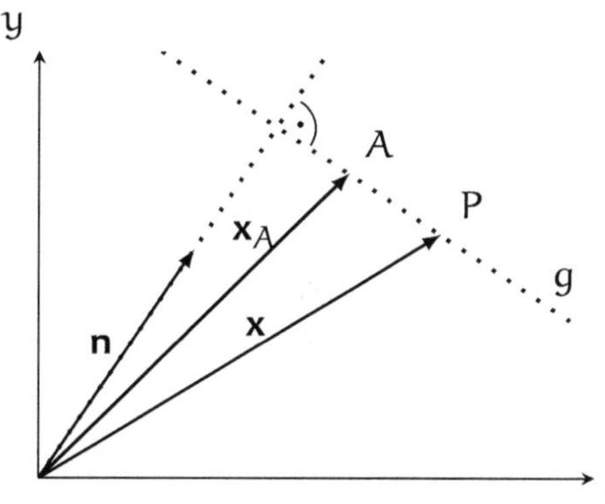

$$(\mathbf{x} - \mathbf{x}_A) \cdot \mathbf{n} = 0$$
$$\mathbf{x} \cdot \mathbf{n} - \mathbf{x}_A \cdot \mathbf{n} = 0$$
$$\mathbf{x} \cdot \mathbf{n} - \text{const.} = 0$$

If **n** is normalized , i.e. a unit normal vector, then the constant is the distance from the line to the origin.

I will explain this in more detail on the next page.

# How can you represent a line in Hesse normal form? (II)

If you normalize the vector $\mathbf{n}$ (i.e. scale it so that it has length 1), then $\mathbf{x} \cdot \mathbf{n_0}$ is the length of the projection of $\mathbf{x}$ onto the direction of $\mathbf{n_0}$, i.e. the distance from $\mathbf{x}$ to the origin line perpendicular to $\mathbf{n}$.

Thus, all points on a line perpendicular to $\mathbf{n}$ with distance d have the value $\mathbf{x} \cdot \mathbf{n_0} = d$.

Thus the expression $d(\mathbf{x}, g) = \mathbf{x} \cdot \mathbf{n_0} - d$ gives the (signed) distance of a point $\mathbf{x}$ from the line g.

If the value of $d(\mathbf{x}, g)$ is zero, the point lies on the line. If the distance is negative, the point lies on the same side of the line as the origin. If the value is positive, it lies on the other side of the line from the origin.

The separation of $\mathbb{R}^2$ or $\mathbb{R}^n$ into two regions is also fundamental for data science.

This representation of a line in the plane using the normal vector is called the Hesse normal form.

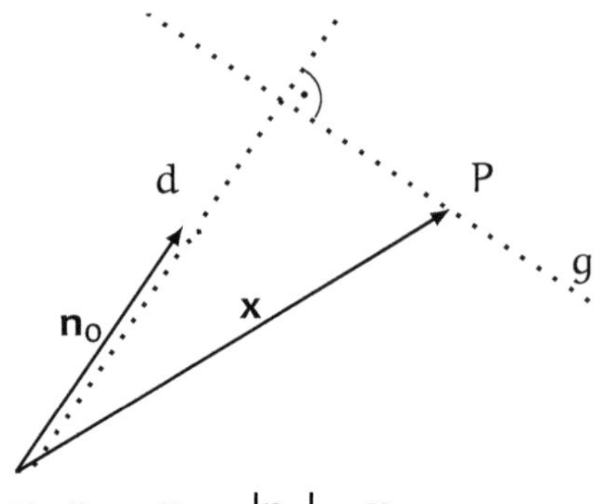


$$\mathbf{x} \cdot \mathbf{n_0} = \mathbf{x}_{\|\mathbf{n_0}} \underbrace{|\mathbf{n_0}|}_{=1} = \mathbf{x}_{\|\mathbf{n_0}}$$

Line g with unit normal vector $\mathbf{n_0}$ and distance $d \geq 0$ from the origin:

$$g: \mathbf{x} \cdot \mathbf{n_0} = d$$
$$g: \mathbf{x} \cdot \mathbf{n_0} - d = 0$$

$$d(\mathbf{x}, g) = \mathbf{x} \cdot \mathbf{n_0} - d$$

$$\begin{cases} < 0 & \mathbf{x} \text{ same side of the origin} \\ = 0 & \mathbf{x} \text{ on the line} \\ > 0 & \mathbf{x} \text{ opposite side of the origin} \end{cases}$$

For the origin, $d(\mathbf{0}, g) = -d \leq 0$.

A perceptron as a simple building block of a neural network can "learn" the separating hyperplane, i.e. $\mathbf{n_0}$ and $d$, from a sample of classified data points.

The Hesse normal form is an implicit equation for the line.

# How can you represent a plane in Hesse normal form?

The Hesse normal form of a plane in three dimensions works completely analogously to that of a line in two dimensions.

The position vectors of all points lying on a plane that is perpendicular to the unit vector $\mathbf{n}_0$ and has a distance d from the origin (measured in the direction of $\mathbf{n}_0$), have the value d as the scalar product with $\mathbf{n}_0$.

plane P with unit normal vector $\mathbf{n}_0$ and distance $d \geq 0$ from the origin:
$$P: \mathbf{x} \cdot \mathbf{n}_0 = d$$
$$P: \mathbf{x} \cdot \mathbf{n}_0 - d = 0$$

If the value of $d(\mathbf{x}, P)$ is zero, the point lies on the plane. If the distance is negative, the point lies on the same side of the plane as the origin. If the value is positive, it lies on the other side of the plane from the origin.

$$d(\mathbf{x}, P) = \mathbf{x} \cdot \mathbf{n}_0 - d$$

$$\begin{cases} < 0 & \mathbf{x} \text{ same side of the origin} \\ = 0 & \mathbf{x} \text{ on the plane} \\ > 0 & \mathbf{x} \text{ opposite side of the origin} \end{cases}$$

If you calculate the scalar product as the sum of the products of the components and move the constant to the other side and normalize to 1, you obtain the plane in intercept form.

The HNF form of the plane with normal vector $(1, 2, 2)^\top$ and distance 2 from the origin,

$$P: \begin{pmatrix} 1/3 \\ 2/3 \\ 2/3 \end{pmatrix} \cdot \mathbf{x} - 2 = 0,$$

yields $x_1 + 2x_2 + 2x_3 = 6$ and thus the intercept form

$$\frac{x_1}{6} + \frac{x_2}{3} + \frac{x_3}{3} = 1.$$

So the plane has the intercepts 6, 3, and 3 with the three axes.

# How can you convert different forms of line or plane equations into each other? (I)

There is an explicit form for the equation of lines in the plane and planes in space, i.e. an equation where one variable shows up only on the left hand side.

This form is very similar to the Hesse normal form. You can easily convert the two forms into each other. To do this, write the linear combination of $x$ and $y$ (and $z$ for planes) as a scalar product. The coefficients of the linear combination become the components of the normal vector. Then you normalize the normal vector and bring all terms to one side to obtain the HNF.

The parametric form of lines in the plane can be converted into HNF by multiplying with a vector that is perpendicular to the direction vector, so that the parameter is eliminated.

For planes, you need a vector that is perpendicular to both direction vectors of the plane. You can easily find this using the cross product, which I will explain later.

$g: y = x + 1$

$P: z = x + y + 1$

$g: -x + y = 1$

$$g: \begin{pmatrix} -1 \\ 1 \end{pmatrix} \cdot \begin{pmatrix} x \\ y \end{pmatrix} = 1$$

$$g: \begin{pmatrix} -\sqrt{2}/2 \\ \sqrt{2}/2 \end{pmatrix} \cdot \begin{pmatrix} x \\ y \end{pmatrix} - \frac{\sqrt{2}}{2} = 0$$

$P: -x - y + z = 1$

$$P: \begin{pmatrix} -1 \\ -1 \\ 1 \end{pmatrix} \cdot \begin{pmatrix} x \\ y \\ z \end{pmatrix} = 1$$

$$P: \begin{pmatrix} -\sqrt{3}/3 \\ -\sqrt{3}/3 \\ \sqrt{3}/3 \end{pmatrix} \cdot \begin{pmatrix} x \\ y \\ z \end{pmatrix} - \frac{\sqrt{3}}{3} = 0$$

$$g: \begin{pmatrix} x \\ y \end{pmatrix} = \begin{pmatrix} 1 \\ 2 \end{pmatrix} + r \begin{pmatrix} 1 \\ -1 \end{pmatrix} \quad | \cdot \begin{pmatrix} 1 \\ 1 \end{pmatrix}$$

$$g: \begin{pmatrix} 1 \\ 1 \end{pmatrix} \cdot \begin{pmatrix} x \\ y \end{pmatrix} = \begin{pmatrix} 1 \\ 1 \end{pmatrix} \cdot \begin{pmatrix} 1 \\ 2 \end{pmatrix} + r \underbrace{\begin{pmatrix} 1 \\ 1 \end{pmatrix} \cdot \begin{pmatrix} 1 \\ -1 \end{pmatrix}}_{=0}$$

$$g: \begin{pmatrix} 1 \\ 1 \end{pmatrix} \cdot \begin{pmatrix} x \\ y \end{pmatrix} = 3$$

$$g: \begin{pmatrix} \sqrt{2}/2 \\ \sqrt{2}/2 \end{pmatrix} \cdot \begin{pmatrix} x \\ y \end{pmatrix} - \frac{3\sqrt{2}}{2} = 0$$

In the last step, I normalized the normal vector and brought everything to one side.

# How can you convert different forms of line or plane equations into each other? (II)

If you have the HNF of a line in 2D and want the parametric form, you solve the linear equation that the HNF directly provides, just as you would solve a linear system with one equation and two variables.

$$g: \begin{pmatrix} \sqrt{2}/2 \\ \sqrt{2}/2 \end{pmatrix} \cdot \begin{pmatrix} x \\ y \end{pmatrix} - \frac{3\sqrt{2}}{2} = 0$$

$g: x + y = 3$

You can choose e.g. $y$ as the parameter:
$y = t, \quad t \in \mathbb{R}.$

It follows that:
$x = 3 - t.$

Thus you obtain:

$$\begin{pmatrix} x \\ y \end{pmatrix} = \begin{pmatrix} 3 \\ 0 \end{pmatrix} + t \begin{pmatrix} -1 \\ 1 \end{pmatrix}, \quad t \in \mathbb{R}.$$

Since the parametric form is not unique, you can obtain different parametric forms.

For the conversion of the HNF of a plane in 3D into the parametric form, you proceed in exactly the same way, except that you now solve a linear system with one equation and three variables.

$$P: \begin{pmatrix} \sqrt{3}/3 \\ \sqrt{3}/3 \\ \sqrt{3}/3 \end{pmatrix} \cdot \begin{pmatrix} x \\ y \\ z \end{pmatrix} - \frac{\sqrt{3}}{3} = 0$$

$P: x + y + z = 1$

You can choose e.g. $y$ and $z$ as parameters:
$y = r, \quad r \in \mathbb{R}$
$z = s, \quad s \in \mathbb{R}.$

It follows that:
$x = 1 - r - s.$

Thus you obtain:

$$\begin{pmatrix} x \\ y \\ z \end{pmatrix} = \begin{pmatrix} 1 \\ 0 \\ 0 \end{pmatrix} + r \begin{pmatrix} -1 \\ 1 \\ 0 \end{pmatrix} + s \begin{pmatrix} -1 \\ 0 \\ 1 \end{pmatrix}$$

$$r, s \in \mathbb{R}.$$

# Overview: Parametric and implicit equations

| | Parametric | Implicit |
|---|---|---|
| **Line in 2d** | $\mathbf{x} = \mathbf{x}_A + r\mathbf{u}, \quad r \in \mathbb{R}$ | $(\mathbf{x} - \mathbf{x}_A) \cdot \mathbf{n} = d$ (HNF) |
| **Line in 3d** | $\mathbf{x} = \mathbf{x}_A + r\mathbf{u}, \quad r \in \mathbb{R}$ | uncommon (would require two equations) |
| **Plane in 3d** | $\mathbf{x} = \mathbf{x}_A + r\mathbf{u} + s\mathbf{v}, \quad r, s \in \mathbb{R}$ | $(\mathbf{x} - \mathbf{x}_A) \cdot \mathbf{n} = d$ (HNF) |
| | The number of parameters is generally the dimension of the object. | The number of equations is generally the codimension of the object (dimension of the space minus dimension of the object). |
| **Parametric form and implicit equation are complementary:** | | |
| | Easy to plot (loop over parameter) | Difficult to plot (need to solve implicit equation) |
| | Difficult to check if a point lies on it (need to solve system of equations) | Easy to check if a point lies on it (by substitution) |
| **An example of nonlinear algebraic objects** that can be described by polynomial equations and are called varieties: | | |
| **Circle in 2d** | Rational (algebraic) parametrization:<br>$$x = \frac{1 - t^2}{1 + t^2}$$ $$y = \frac{2t}{1 + t^2}$$ $t \in \mathbb{R} \cup \{\infty\}$ <br><br>Trigonometric (transcendental) parametrization:<br>$$x = \cos(\varphi)$$ $$y = \sin(\varphi)$$ $\varphi \in [0, 2\pi)$ | $x^2 + y^2 = 1$ |

Explicit equations like $z = 2x + y$ are only possible if "above every point" of the space of the independent variables (essentially the parameters, here $x$ and $y$) there is at most one point of the object, since the object is represented as the graph of a function of this explicit equation.

# Vectors, Matrices, and Systems of Linear Equations

Vectors are not only relevant in a 2-dimensional geometric space and the 3-dimensional physical space we inhabit, but they are also a very powerful concept in many areas of mathematics.

I will explain what general vector spaces, also known as linear spaces, are, as well as what a basis of a vector space signifies.

On vector spaces you can define linear mappings. These are closely related to systems of linear equations, whose solution set forms a vector space or an affine space.

Linear mappings can be represented as matrices, which are two-dimensional arrays of numbers. For square matrices, you can calculate the determinant, which is a number that characterizes the matrix.

With the help of the determinant, we can define another type of product between two vectors in three dimensions: the cross product, also called vector product.

Finally, I will bring together the chapter on analytic geometry and the chapter on linear algebra, and show you how to use systems of linear equations to determine the intersections of affine subspaces, such as lines and planes.



A. Gründers, *Math Made Clear: From the Basics to Calculus*, https://doi.org/10.1007/978-3-662-73221-2_12

# What is an (abstract) vector space?

A vector space over the real numbers is first a set $V$ in which you can add and subtract just as in the integers. This is called a group.

$(V, +)$ forms an abelian group:

(i) For all $\mathbf{v}, \mathbf{w} \in V$ holds:
$\mathbf{v} + \mathbf{w} = \mathbf{w} + \mathbf{v} \in V$.

(ii) There exists $\mathbf{0} \in V$ (zero vector), such that $\mathbf{v} + \mathbf{0} = \mathbf{v}$.

(iii) For every $\mathbf{v} \in V$ there exists $-\mathbf{v} \in V$ such that $\mathbf{v} + (-\mathbf{v}) = \mathbf{0}$.

Examples of abelian groups are $(\mathbb{Z}, +), (\mathbb{Q}, +), (\mathbb{R}, +), (\mathbb{R}_{>0}, \cdot)$.

$(\mathbb{N}, +)$ is not an abelian group, since 0 is not included, and there is also no element that can be added to 1 to yield 0.

In addition, you can multiply the vectors by a real number and obtain another vector. In this context, the numbers are also called scalars, and the multiplication is called scalar multiplication.

For all $\mathbf{v} \in V$ and $r \in \mathbb{R}$:
$r \cdot \mathbf{v} \in V$, also written as $r\mathbf{v}$.

Then $(V, +, \cdot)$ is a real vector space, also called a vector space over the real numbers.

For a vector space over the complex numbers, the following holds: For all $\mathbf{v} \in V$ and $r \in \mathbb{C}$, $r\mathbf{v} \in V$. You can also use other number fields as a basis, such as the rational numbers $\mathbb{Q}$; but it's important that you can subtract and divide in them as in the real numbers, so that they form a field.

The scalar multiplication must be compatible with addition, which is ensured, among other things, by the distributive laws.

$r(\mathbf{v} + \mathbf{w}) = r\mathbf{v} + r\mathbf{w}$
$(r + s)\mathbf{v} = r\mathbf{v} + s\mathbf{v}$
$(rs)\mathbf{v} = r(s\mathbf{v})$
$1\mathbf{v} = \mathbf{v}$

Examples of real vector spaces are the set of 2-dimensional real vectors and the set of real sequences, each with componentwise addition and multiplication by a real number.

It holds:
$\mathbf{v} + \mathbf{w} \in \mathbb{R}^2$ for $\mathbf{v}, \mathbf{w} \in \mathbb{R}^2$ and
$r\mathbf{v} \in \mathbb{R}^2$ for $r \in \mathbb{R}, \mathbf{v} \in \mathbb{R}^2$.

The sum of two real sequences is again a real sequence, and the product of a real sequence with a real number is also a real sequence.

# What is basis and dimension of a vector space?

If you form all possible linear combinations of a set of vectors, you obtain the so-called span of these vectors. For example, two linearly independent vectors span a plane.

If for a set of vectors of a given vector space the span is equal to the entire vector space, then we say that this set of vectors generates the vector space.

A linearly independent set of vectors of $V$ that generates the entire vector space $V$ is called a basis of $V$.

It can be shown that all bases of a vector space have the same cardinality. The cardinality of a basis is called the dimension of the vector space considered.

In a finite-dimensional vector space $V$ of dimension $n$, $n$ linearly independent vectors form a basis, and $n$ vectors that generate $V$ also form a basis.

Thus, every vector can be written uniquely as a finite linear combination (i.e. a weighted sum) of basis vectors.

$$\text{span}\,(\mathbf{a}_1, \mathbf{a}_2, \ldots \mathbf{a}_n)$$
$$= \{r_1\mathbf{a}_1 + r_2\mathbf{a}_2 + \ldots + r_n\mathbf{a}_n, r_i \in \mathbb{R}\}$$

Note that by definition, a linear combination always consists of only finitely many summands, even if the vector space is infinite-dimensional.

$$\text{span}\,((0,1)^\top,(1,0)^\top) = \mathbb{R}^2$$
$$\Rightarrow \{(0,1)^\top,(1,0)^\top\} \text{ generates } \mathbb{R}^2.$$

$$\text{span}\,((0,1)^\top,(0,2)^\top) \neq \mathbb{R}^2$$
$$\Rightarrow \{(0,1)^\top,(0,2)^\top\} \text{ doesn't generate } \mathbb{R}^2.$$

$$\text{span}\,((0,1)^\top,(1,0)^\top) = \mathbb{R}^2$$
$$\{(0,1)^\top,(1,0)^\top\} \text{ is a basis.}$$

$$\text{span}\,((0,1)^\top,(1,0)^\top,(1,1)^\top) = \mathbb{R}^2$$
$$\{(0,1)^\top,(1,0)^\top,(1,1)^\top\} \text{ is not a basis.}$$

Moreover, it can be shown that every finite-dimensional vector space over $\mathbb{R}$ is isomorphic (i.e. identical up to the naming of the elements) to some $\mathbb{R}^n, n \in \mathbb{N}$.

$n$ linearly independent vectors thus "automatically" generate $V$.

$n$ vectors that generate $V$ are thus "automatically" lin. independent.

From now on, we will only consider finite-dimensional vector spaces. An infinite-dimensional vector space is, for example, the vector space of all real sequences.

# What is a linear map?

A map between vector spaces is a linear map, if it does not matter whether you first form a linear combination and then apply the map, or first apply the map and then form the linear combination.

$f : V \to W$ is a linear map, if

$$\underbrace{f(r\mathbf{v} + s\mathbf{w})}_{\text{first linear comb.}} = \underbrace{rf(\mathbf{v}) + sf(\mathbf{w})}_{\text{first apply map}}$$

for all $r, s \in \mathbb{R}; \mathbf{v}, \mathbf{w} \in V$.

This means the map is compatible with linear combinations.
This is equivalent to:

$f(\mathbf{v} + \mathbf{w}) = f(\mathbf{v}) + f(\mathbf{w})$ for all $v, w \in V$
$f(r\mathbf{v}) = rf(\mathbf{v})$ for all $r \in \mathbb{R}, v \in V$.

A rotation around the origin is a linear map from the plane to itself.

$f : \mathbb{R}^2 \to \mathbb{R}^2$

$$\begin{pmatrix} x \\ y \end{pmatrix} \mapsto \begin{pmatrix} -y \\ x \end{pmatrix}$$

describes a rotation by $90°$ counterclockwise and is compatible with linear combinations, so it's a linear map.

The addition of a constant vector is not a linear map.

$f : \mathbb{R}^2 \to \mathbb{R}^2$

$$\begin{pmatrix} x \\ y \end{pmatrix} \mapsto \begin{pmatrix} x \\ y \end{pmatrix} + \begin{pmatrix} 1 \\ 0 \end{pmatrix} = \begin{pmatrix} x + 1 \\ y \end{pmatrix}$$

is not linear, because

$$f(2v) = f\begin{pmatrix} 2x \\ 2y \end{pmatrix} = \begin{pmatrix} 2x + 1 \\ 2y \end{pmatrix}$$

$$\neq \begin{pmatrix} 2x + 2 \\ 2y \end{pmatrix} = 2\begin{pmatrix} x + 1 \\ y \end{pmatrix} = 2f(v)$$

The derivative on the vector space of differentiable functions is a linear map.

$$\frac{d}{dx}(af(x) + bg(x)) = a\frac{df(x)}{dx} + b\frac{dg(x)}{dx}$$

# What do you need matrices for?

With matrices you can describe linear mappings $f : V \to W$.

$$v \xrightarrow{\ f\ } f(v)$$
$$\in V \qquad \in W$$

Since every vector space has a basis, you can choose a basis in $V$. You can represent every vector from $V$ in this basis.

$$V = \langle e_1, e_2, \ldots, e_n \rangle, \ n = \dim(V)$$

$$v = v_1 e_1 + \cdots + v_n e_n, \ v \in V$$

Since a linear map is compatible with linear combinations, it's enough to specify the linear map $f$ for these basis vectors.

$$f(v) = v_1 f(e_1) + \cdots + v_n f(e_n)$$
$$= \sum_{i=1}^{n} v_i f(e_i) \quad (*)$$

You can also choose a basis in $W$. If you expand $f(e_i)$ in this basis, the coefficients depend on two indices.

$$W = \langle b_1, b_2, \ldots, b_m \rangle, \ m = \dim(W)$$

$$f(e_i) = a_{1i} b_1 + a_{2i} b_2 + \cdots + a_{mi} b_m$$

You can arrange the $mn$ coefficients $a_{ji}$ with $j = 1, \ldots, m$, $i = 1, \ldots, n$ in a matrix. The $i$-th column shows you the components of the image of the vector $e_i, i = 1, \ldots, n$, in the basis $b_j, j = 1, \ldots, m$.

$$\begin{pmatrix} a_{11} & \cdots & a_{1i} & \cdots & a_{1n} \\ a_{21} & \cdots & a_{2i} & \cdots & a_{2n} \\ \vdots & & \vdots & & \vdots \\ a_{m1} & \cdots & a_{mi} & \cdots & a_{mn} \end{pmatrix}$$

$$\uparrow$$

$$f(e_i) = \sum_{j=1}^{m} a_{ji} b_j \quad (**)$$

If you substitute $(**)$ into $(*)$, you find the representation of $f(v)$ using the matrix elements.

$$f(v) = \sum_{i=1}^{n} \sum_{j=1}^{m} a_{ji} v_i b_i = \sum_{j=1}^{n} f(v)_j b_j$$

$$\text{with } f(v)_j = \sum_{i=1}^{n} a_{ji} v_i$$

This $m \times n$ matrix $M \in \mathbb{R}^{m \times n}$ describes the linear mapping $f : V \cong \mathbb{R}^n \to W \cong \mathbb{R}^m$. The number of columns $n$ is the dimension of the vector space $V$, the number of rows $m$ is the dimension of the vector space $W$.

The $j$-th component $f(v)_j$ of the image vector of $v$ is the scalar product of the $j$-th row vector $(a_{ji})_{i=1,\ldots,n}$ with the column vector $v = (v_i)_{i=1,\ldots,n}$.

# How do you add matrices and multiply them by a scalar?

You add two matrices just like two vectors, that is, element by element. Therefore, the two matrices must be of the same type, i.e. have matching numbers of rows and columns.

$$\begin{pmatrix} 1 & 2 \\ 2 & 3 \\ 3 & 4 \end{pmatrix} + \begin{pmatrix} 1 & 0 \\ 0 & -1 \\ 1 & 1 \end{pmatrix} = \begin{pmatrix} 2 & 2 \\ 2 & 2 \\ 4 & 5 \end{pmatrix}$$

You multiply a matrix by a number (scalar) just as you do with a vector, by multiplying each element by the number.

$$2 \begin{pmatrix} 1 & 2 \\ 2 & 3 \\ 3 & 4 \end{pmatrix} = \begin{pmatrix} 2 & 4 \\ 4 & 6 \\ 6 & 8 \end{pmatrix}$$

Matrices can be transposed, which means swapping columns and rows. Visually, you reflect the entries across the diagonal running from the top left to the bottom right.

$$\begin{pmatrix} 1 & 2 & 3 \\ 4 & 5 & 6 \end{pmatrix}^{\top} = \begin{pmatrix} 1 & 4 \\ 2 & 5 \\ 3 & 6 \end{pmatrix}$$

If you transpose a matrix twice, you get the original matrix back.

$$\left( \mathbf{A}^{\top} \right)^{\top} = \mathbf{A}$$

These operations are compatible with each other: it doesn't matter the order in which you perform them. For example, you get the same result whether you first add matrices and then multiply by a number, or first multiply each matrix by the number and then add the results.

$$r(\mathbf{A} + \mathbf{B}) = r\mathbf{A} + r\mathbf{B}$$

$$(\mathbf{A} + \mathbf{B})^{\top} = \mathbf{A}^{\top} + \mathbf{B}^{\top}$$

$$(r\mathbf{A})^{\top} = r\mathbf{A}^{\top}$$

However, when multiplying two matrices together, the order is important.

I will explain matrix multiplication to you on the next page.

# How do you multiply matrices?

A matrix represents a linear mapping. When you perform two linear maps one after the other, this corresponds to multiplying the respective two matrices.

For the case of two $2 \times 2$ matrices, you can see the calculation on the right hand side.

$$x \xrightarrow{f_B} Bx \xrightarrow{f_A} A(Bx)$$

$$x \xrightarrow{f_{AB}=f_A \circ f_B} (AB)x$$

$$B = \begin{pmatrix} b_{11} & b_{12} \\ b_{21} & b_{22} \end{pmatrix}, \quad x = \begin{pmatrix} x_1 \\ x_2 \end{pmatrix}$$

$$Bx = \begin{pmatrix} b_{11}x_1 + b_{12}x_2 \\ b_{21}x_1 + b_{22}x_2 \end{pmatrix} = \begin{pmatrix} (Bx)_1 \\ (Bx)_2 \end{pmatrix}$$

$$A(Bx) = \begin{pmatrix} a_{11}(Bx)_1 + a_{12}(Bx)_2 \\ a_{21}(Bx)_1 + a_{22}(Bx)_2 \end{pmatrix}$$

$$= \begin{pmatrix} a_{11}(b_{11}x_1 + b_{12}x_2) + a_{12}(b_{21}x_1 + b_{22}x_2) \\ a_{21}(b_{11}x_1 + b_{12}x_2) + a_{22}(b_{21}x_1 + b_{22}x_2) \end{pmatrix}$$

$$= \begin{pmatrix} (a_{11}b_{11}+a_{12}b_{21})x_1 + (a_{11}b_{12}+a_{12}b_{22})x_2 \\ (a_{21}b_{11}+a_{22}b_{21})x_1 + (a_{21}b_{12}+a_{22}b_{22})x_2 \end{pmatrix}$$

Because $A(Bx) = (AB)x$,

$$AB = \begin{pmatrix} a_{11}b_{11} + a_{12}b_{21} & a_{11}b_{12} + a_{12}b_{22} \\ a_{21}b_{11} + a_{22}b_{21} & a_{21}b_{12} + a_{22}b_{22} \end{pmatrix}.$$

For example the $(1,1)$-element of the resulting matrix is the sum of the products $a_{1i}b_{i1}$ for all $i$. This formula can be generalized from the $(1,1)$-element to all elements.

$$(AB)_{11} = a_{11}b_{11} + a_{12}b_{21} = \sum_{i=1}^{2} a_{1i}b_{i1}$$

The $(r,s)$-element is the sum of the products $a_{ri}b_{is}$ over $i = 1, \ldots, n$, i.e. the scalar product of the $r$-th row of $A$ with the $s$-th column of $B$.

$$(AB)_{rs} = \sum_{i=1}^{n} a_{ri}b_{is}$$

$$\begin{pmatrix} & \cdots & b_{1s} & \cdots \\ & & \vdots & \\ & \cdots & b_{ns} & \cdots \end{pmatrix}$$

$$\begin{pmatrix} \vdots & & \vdots \\ a_{r1} & \cdots & a_{rn} \\ \vdots & & \vdots \end{pmatrix} \begin{pmatrix} \vdots \\ \cdots & (AB)_{rs} & \cdots \\ \vdots \end{pmatrix}$$

$$(AB)_{rs} = a_{r1}b_{1s} + \cdots + a_{rn}b_{ns}$$

# How do you concretely multiply matrices?

First, you check whether you can multiply the two matrices. For this, the number of columns of the matrix on the left must be equal to the number of rows of the matrix on the right.

If you multiply a $(r_1 \times c_1)$ matrix from the left with a $(r_2 \times c_2)$ matrix, then $c_1 = r_2$ must hold, otherwise matrix multiplication is not possible:

$(1 \times 3)(3 \times 2)$ is possible.
$(1 \times 3)(2 \times 3)$ is not possible.

The easiest way is to write the matrices as shown on the right.

$$\begin{pmatrix} 4 & 5 & 6 \\ 7 & 8 & 9 \end{pmatrix}$$
$$\begin{pmatrix} 1 & 2 \\ 3 & 4 \end{pmatrix} \begin{pmatrix} 18 & 21 & 24 \\ 40 & 47 & 54 \end{pmatrix}$$

Each element of the matrix product is obtained as the scalar product of the corresponding row vector of the first matrix with the corresponding column vector of the second matrix, i.e. you multiply the corresponding elements and sum them up.

For example, the element at the bottom left is

$3 \cdot 4 + 4 \cdot 7 = 12 + 28 = 40.$

Matrix multiplication is generally not commutative, i.e. the order matters.

$$\begin{pmatrix} 1 & 1 \\ 0 & 0 \end{pmatrix} \begin{pmatrix} 1 & 0 \\ 1 & 0 \end{pmatrix} = \begin{pmatrix} 2 & 0 \\ 0 & 0 \end{pmatrix}$$

$$\begin{pmatrix} 1 & 0 \\ 1 & 0 \end{pmatrix} \begin{pmatrix} 1 & 1 \\ 0 & 0 \end{pmatrix} = \begin{pmatrix} 1 & 1 \\ 1 & 1 \end{pmatrix}$$

If you multiply by the identity matrix, the matrix does not change.

$$\begin{pmatrix} 1 & 0 \\ 0 & 1 \end{pmatrix} \begin{pmatrix} a & b \\ c & d \end{pmatrix} = \begin{pmatrix} a & b \\ c & d \end{pmatrix}$$

If you multiply by the so-called inverse matrix, you get the identity matrix.

$$\begin{pmatrix} 1 & 2 \\ 3 & 4 \end{pmatrix} \begin{pmatrix} -2 & 1 \\ 3/2 & -1/2 \end{pmatrix} = \begin{pmatrix} 1 & 0 \\ 0 & 1 \end{pmatrix}$$

In this case of an inverse matrix, it holds true regardless of the order.

$$\begin{pmatrix} -2 & 1 \\ 3/2 & -1/2 \end{pmatrix} \begin{pmatrix} 1 & 2 \\ 3 & 4 \end{pmatrix} = \begin{pmatrix} 1 & 0 \\ 0 & 1 \end{pmatrix}$$

# What are important special cases of matrix multiplication?

Multiplying an $n \times k$ matrix with a $k \times m$ matrix yields an $n \times m$ matrix:

$$(n \times k)(k \times m) = (n \times m).$$

Each element (black) of the product matrix (black framed) is the scalar product of the corresponding row vector of the $n \times k$ matrix (gray) with the corresponding column vector of the $k \times m$ matrix (gray).

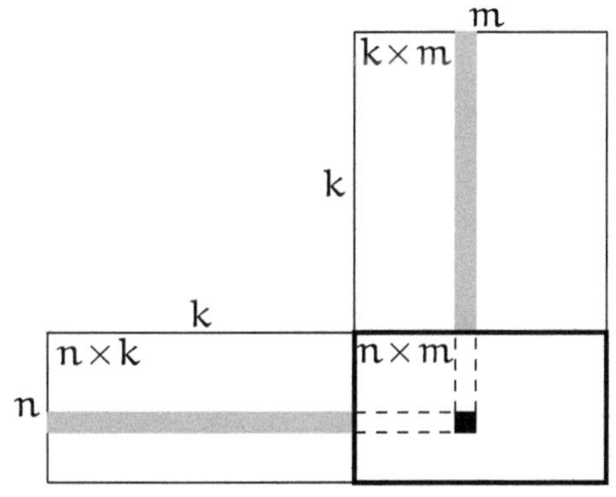

Important special cases are:

$(n \times n)(n \times 1) = (n \times 1)$:
Multiplying a square matrix by a vector yields a vector of the same dimension, e.g. the rotated vector when multiplying by a rotation matrix.

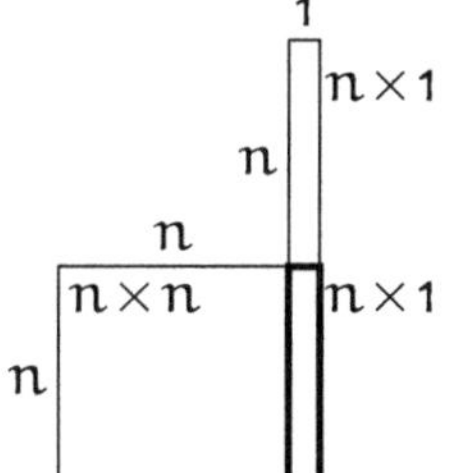

$A \in \mathbb{R}^{n \times n}$
$b \in \mathbb{R}^n \cong \mathbb{R}^{n \times 1}$
$Ab \in \mathbb{R}^n$
$(Ab)_i = \sum_{j=1}^{n} A_{ij} b_j$

$(1 \times n)(n \times 1) = (1 \times 1)$:
Scalar product: The matrix multiplication of a row vector with a column vector yields a matrix with only one row and one column, i.e. a number (=scalar).

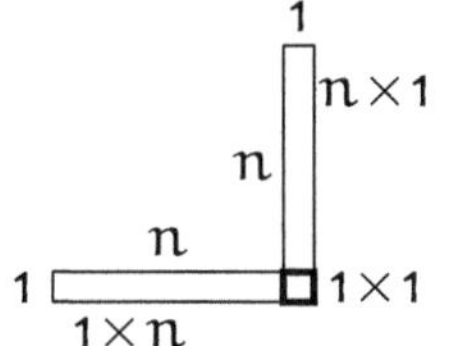

$a, b \in \mathbb{R}^n \cong \mathbb{R}^{n \times 1}$
$a^{\top} \in \mathbb{R}^{1 \times n}$
$a^{\top} b = a \cdot b \in \mathbb{R}$
$a^{\top} b = \sum_{i=1}^{n} a_i b_i$

$(n \times 1)(1 \times n) = (n \times n)$:
Dyadic product: The matrix multiplication of a column vector with a row vector yields a square matrix of the dimension of the vectors.

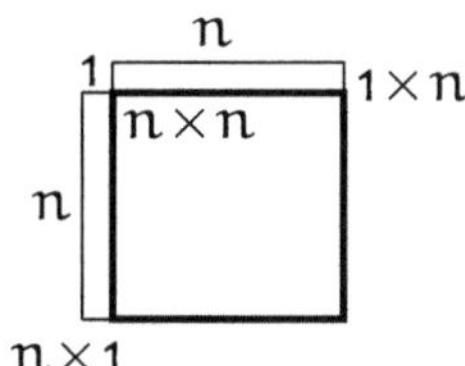

$a, b \in \mathbb{R}^n \cong \mathbb{R}^{n \times 1}$
$b^{\top} \in \mathbb{R}^{1 \times n}$
$ab^{\top} \in \mathbb{R}^{n \times n}$
$(ab^{\top})_{ij} = a_i b_j$

# What is the inverse matrix?

We only consider square matrices in the sequel.

The inverse matrix is the matrix which, when multiplied by a given matrix, yields the identity matrix.

You can show that if a matrix is the inverse of another matrix with respect to the multiplication from the left, it is also the inverse with respect to the multiplication from the right.

Not every matrix has an inverse matrix, but only those matrices that have full rank or – equivalently – a nonzero determinant. These matrices are called invertible.

Calculating the inverse matrix is equivalent to solving a system of linear equations.

For invertible $2 \times 2$ matrices, you can easily write down the inverse.

With the inverse, you can solve a system of linear equations.

If you know that the inverse exists, then it is easy to see that the unique solution of the homogeneous system is 0, i.e. there is only the trivial solution.

$$\begin{pmatrix} 1 & 2 \\ 3 & 4 \end{pmatrix}^{-1} = \begin{pmatrix} -2 & 1 \\ 3/2 & -1/2 \end{pmatrix}, \text{ because}$$

$$\begin{pmatrix} -2 & 1 \\ 3/2 & -1/2 \end{pmatrix}\begin{pmatrix} 1 & 2 \\ 3 & 4 \end{pmatrix} = \begin{pmatrix} 1 & 0 \\ 0 & 1 \end{pmatrix}.$$

$$\begin{pmatrix} 1 & 2 \\ 3 & 4 \end{pmatrix}\begin{pmatrix} -2 & 1 \\ 3/2 & -1/2 \end{pmatrix} = \begin{pmatrix} 1 & 0 \\ 0 & 1 \end{pmatrix}$$

$$\begin{pmatrix} 1 & 2 \\ 2 & 4 \end{pmatrix} \text{ has no inverse.}$$

Its determinant is 0, its rank is 1.
Full rank means that the rank is maximal, i.e. for a square matrix, equal to the number of rows = number of columns.

I will show you this when solving systems of linear equations.

$$\begin{pmatrix} a & b \\ c & d \end{pmatrix}^{-1} = \frac{1}{ad - bc}\begin{pmatrix} d & -b \\ -c & a \end{pmatrix}$$

There is also a similar formula for $n \times n$ matrices. If you are interested, look up "adjugate" on the internet.

$$A\mathbf{x} = \mathbf{b} \Rightarrow \mathbf{x} = A^{-1}\mathbf{b}$$

However, the Gaussian algorithm is more efficient.

$$A\mathbf{x} = \mathbf{0} \Rightarrow \mathbf{x} = A^{-1}\mathbf{0} = \mathbf{0}$$

# What holds for matrix multiplication?

For matrix multiplication, the associative law holds, but not the commutative law.

$$A(BC) = (AB)C$$

$$AB \neq BA \text{ in general}$$

Matrix multiplication is compatible with multiplication by a scalar and with addition.

$$A(rB) = (rA)B = r(AB)$$
$$A(B + C) = AB + AC$$
$$(A + B)C = AC + BC$$

Thus it's a linear operation.

$$A(rB + sC) = rAB + sAC$$

When calculating the inverse of a matrix product, you need to reverse the order of the product.

$$(AB)^{-1} = B^{-1}A^{-1}$$

That the reversal is necessary follows directly as you see from the calculation in the the right hand side and from the fact that matrix multiplication is generally not commutative.

$$(B^{-1}A^{-1})(AB) = B^{-1}\underbrace{A^{-1}A}_{=1}B$$
$$= B^{-1}B = 1$$
$$B^{-1}A^{-1} \neq A^{-1}B^{-1} \text{ in general.}$$

The reversal of the order when inverting applies to all operations that do not commute.

This also holds for inverse functions:
$$(f(g(x)))^{-1} = g^{-1}(f^{-1}(x))$$
or in short $(f \circ g)^{-1} = g^{-1} \circ f^{-1}$

This also applies in everyday life: You put on your socks first and then your shoes. When inverting = undoing = taking off, it's the other way around: You take off your shoes first, then your socks.

The reversal of the order also applies when transposing the product of two matrices.

$$(AB)^{\top} = B^{\top}A^{\top}$$

For example,
$$\mathbf{x}^{\top}A^{\top}A\mathbf{x} = (A\mathbf{x})^{\top}A\mathbf{x}$$
$$= (A\mathbf{x}) \cdot (A\mathbf{x})$$
$$= \|A\mathbf{x}\|^2 \geq 0 \text{ for all } A, \mathbf{x}.$$

# What is a determinant?

Determinants are numbers that characterize square matrices. I will explain to you on the following pages, how to calculate them and what they are good for.

They are written as $\det(M)$ or also like a matrix but with straight lines instead of brackets.

Often it's useful to focus on the dependence on the column vectors. The determinant is then written as a function of the $n$ column vectors, as shown on the right for $n = 3$.

$$\det(A) = \det \begin{pmatrix} a_{11} & \cdots & a_{1n} \\ \vdots & \ddots & \vdots \\ a_{n1} & \cdots & a_{nn} \end{pmatrix} \in \mathbb{R}$$

$$\det(A) = \begin{vmatrix} a_{11} & \cdots & a_{1n} \\ \vdots & \ddots & \vdots \\ a_{n1} & \cdots & a_{nn} \end{vmatrix}$$

$$\det(\mathbf{a}, \mathbf{b}, \mathbf{c}) = \begin{vmatrix} a_1 & b_1 & c_1 \\ a_2 & b_2 & c_2 \\ a_3 & b_3 & c_3 \end{vmatrix}$$

The determinant gives the oriented area or the oriented volume of the parallelogram respectively parallelepiped spanned by the column vectors.

Here, "oriented" volume means that it is positive if the order of the column vectors in 2 dimensions is counterclockwise, or in 3 dimensions as the thumb, index finger, and middle finger of the right hand are oriented, and otherwise negative.

The determinant is multiplicative, but it is not additive. If you swap rows and columns, it remains the same.

$$\det(AB) = \det(A)\det(B)$$
$$\det(A + B) \neq \det(A) + \det(B)$$
$$\det(A^{\top}) = \det(A)$$

When multiplying an $n \times n$ matrix by $a$, the determinant is multiplied by $a^n$.

$$\det(rA) = r^n \det(A)$$
When scaling the vectors by a factor $r$, the $n$-dimensional volume changes by a factor of $r^n$.

The determinant is zero if and only if the columns are linearly dependent. This also holds equally for the rows.

In this sense, the determinant measures the deviation of the set of column vectors from linear dependence.
If the determinant of a matrix is nonzero, the underlying matrix is invertible.

# How can you understand $2 \times 2$ determinants?

The $2 \times 2$ determinant is a number that characterizes a $2 \times 2$ matrix and is calculated from its elements.

It measures the deviation from linear independence of the two column vectors: the determinant is zero if and only if the two columns are linearly dependent. On the right hand side, one direction of the proof is shown, even though not in full generality.

The determinant measures the (oriented) area of the parallelogram spanned by the two column vectors.

On the right, you see that the two gray areas in the upper figure together are equal in size to the gray rectangle in the lower figure. This is because the total rectangle is the same size in both cases, and you can see that the white areas in both figures are equal by shifting them. Therefore, the gray areas are also equal in each case.

Thus the area of the parallelogram spanned by $\begin{pmatrix} a \\ b \end{pmatrix}$ and $\begin{pmatrix} c \\ d \end{pmatrix}$ in the upper figure equals the difference between the areas of the gray rectangle in the lower figure and the gray one in the upper figure.

$$\begin{vmatrix} a & c \\ b & d \end{vmatrix} = ad - bc$$

$$\begin{pmatrix} a \\ b \end{pmatrix} = r \begin{pmatrix} c \\ d \end{pmatrix}, \quad r \in \mathbb{R}$$

$$\Rightarrow r = \frac{a}{c} = \frac{b}{d} \quad \text{(if } c, d \neq 0)$$

$$\Rightarrow ad - bc = 0$$

$\begin{vmatrix} a & c \\ b & d \end{vmatrix}$ measures the area of the parallelogram with sides $\begin{pmatrix} a \\ b \end{pmatrix}$ and $\begin{pmatrix} c \\ d \end{pmatrix}$.

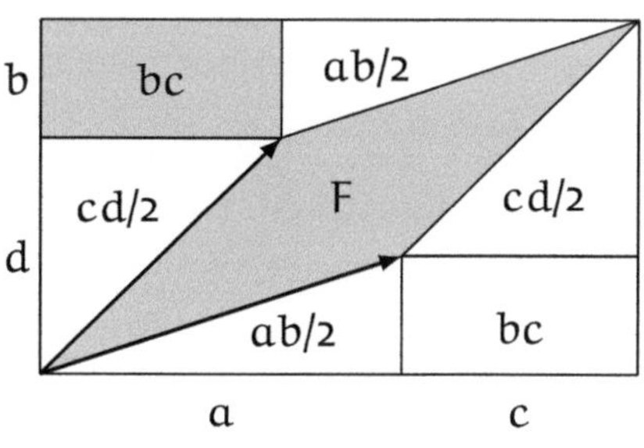


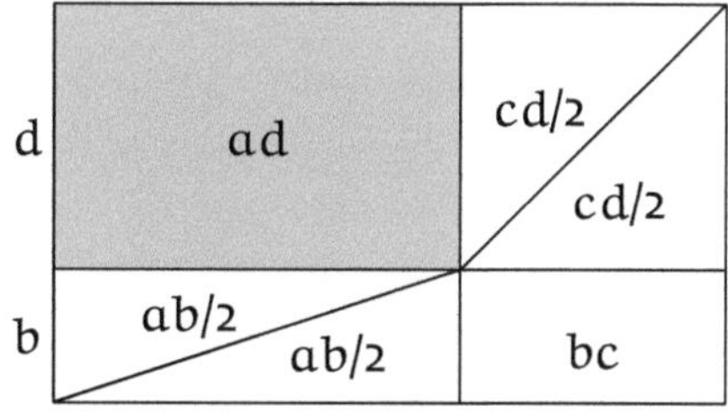


gray areas above
= gray area below:
$$\Rightarrow bc + F = ad$$
$$\Rightarrow F = ad - bc = \begin{vmatrix} a & c \\ b & d \end{vmatrix}$$

# What properties does a determinant have due to its geometric meaning?

A determinant maps $n$ vectors from $\mathbb{R}^n$ to a real number that gives the $n$-dimensional volume of the body spanned by the $n$ vectors .

$$\det : \mathbb{R}^n \times \mathbb{R}^n \times \ldots \times \mathbb{R}^n \to \mathbb{R}$$
$$(\mathbf{v}_1, \mathbf{v}_2, \ldots, \mathbf{v}_n) \mapsto \det(\mathbf{v}_1, \mathbf{v}_2, \ldots, \mathbf{v}_n)$$

In two dimensions: parallelogram, from three dimensions on, it is called a parallelepiped.

If all vectors are unit vectors, the absolute value of the determinant is one. Here we consider $n = 2$, but the properties hold similarly for $n > 2$.

$$\det(\mathbf{e}_1, \mathbf{e}_2) = 1$$
$$\begin{vmatrix} 1 & 0 \\ 0 & 1 \end{vmatrix} = 1 \cdot 1 - 0 \cdot 0 = 1$$

The unit square has area 1.

A determinant changes its sign when you swap two vectors.

$$\det(\mathbf{b}, \mathbf{a}) = -\det(\mathbf{a}, \mathbf{b})$$
$$\begin{vmatrix} c & a \\ d & b \end{vmatrix} = bc - ad = - \begin{vmatrix} a & c \\ b & d \end{vmatrix}$$

Thus, the determinant of two identical vectors is zero.

$$\det(\mathbf{a}, \mathbf{a}) = -\det(\mathbf{a}, \mathbf{a}) \Rightarrow \det(\mathbf{a}, \mathbf{a}) = 0.$$
Two identical vectors do not span an area.

If you multiply one of the vectors by $a$, the determinant is also multiplied by $a$. In addition, the determinant is additive in each argument, since the areas add up.

This results in linearity e.g. in the first argument:
$$\det(r\mathbf{a} + s\mathbf{b}, \mathbf{c}) = r \det(\mathbf{a}, \mathbf{c}) + s \det(\mathbf{b}, \mathbf{c}).$$

From this and with $\det(\mathbf{a}, \mathbf{a}) = 0$, it follows that you can add any multiple of the first vector to the second vector without changing the determinant.

$$\det(\mathbf{e}_1, \mathbf{e}_2 + 2\mathbf{e}_1) = \det(\mathbf{e}_1, \mathbf{e}_2)$$

In a shear transformation, the area remains the same: the parallelogram spanned by $\mathbf{e}_1$ and $\mathbf{e}_2 + 2\mathbf{e}_1$ has the same area as the unit square spanned by $\mathbf{e}_1$ and $\mathbf{e}_2$.

A determinant is called an alternating multilinear form. Alternating means that the sign changes when two vectors are swapped.

A linear form is a linear map that maps a vector to a number.
A multilinear form is a generalization that maps several vectors to a number.

# How can you understand the formula for $3 \times 3$ determinants?

The determinant, here abbreviated as $A$, is uniquely determined by the properties of alternating multilinear forms and the requirement that a unit cube has volume 1 (normalization).

This also holds for $3 \times 3$ determinants for all three arguments.

With this, we can calculate any $3 \times 3$ determinant.

With the multilinearity, we first get all combinations of $A(e_i, e_j, e_k)$ with $i, j, k \in \{1, 2, 3\}$, i.e. 27 terms. Due to (III), only 6 of them are nonzero, namely those where all indices are different, i.e. a permutation of $\{1, 2, 3\}$.

Thus, using (I) we get a sum of 6 terms.

With $A(e_1, e_2, e_3) = 1$ (normalization) and (III), 3 of the 6 terms (which require an even number of swaps) get a positive sign, and the 3 remaining ones get a negative sign.

Here you see it for $2 \times 2$:

(I) $A(a\mathbf{u}, \mathbf{v}) = aA(\mathbf{u}, \mathbf{v})$
(II) $A(\mathbf{u}_1 + \mathbf{u}_2, \mathbf{v}) = A(\mathbf{u}_1, \mathbf{v}) + A(\mathbf{u}_2, \mathbf{v})$
(III) $A(\mathbf{v}, \mathbf{u}) = -A(\mathbf{u}, \mathbf{v})$

(I),(II) for all arguments: multilinear
(III): alternating.

E.g. for the second and third argument in (III):
$A(\mathbf{u}, \mathbf{v}, \mathbf{w}) = -A(\mathbf{u}, \mathbf{w}, \mathbf{v})$
$A(\mathbf{a}, \mathbf{b}, \mathbf{c}) = A(a_1\mathbf{e}_1 + a_2\mathbf{e}_2 + a_3\mathbf{e}_3,$
$b_1\mathbf{e}_1 + b_2\mathbf{e}_2 + b_3\mathbf{e}_3, c_1\mathbf{e}_1 + c_2\mathbf{e}_2 + c_3\mathbf{e}_3)$

If two indices are equal, the value is zero. E.g. from (III), for the second and third argument, it follows for the term with $\mathbf{e}_1, \mathbf{e}_3, \mathbf{e}_3$:
$A(\mathbf{e}_1, \mathbf{e}_3, \mathbf{e}_3) = -A(\mathbf{e}_1, \mathbf{e}_1, \mathbf{e}_3)$
$\Leftrightarrow A(\mathbf{e}_1, \mathbf{e}_3, \mathbf{e}_3) = 0.$

$A(\mathbf{a}, \mathbf{b}, \mathbf{c}) =$
$a_1 b_2 c_3 A(\mathbf{e}_1, \mathbf{e}_2, \mathbf{e}_3) + a_1 b_3 c_2 A(\mathbf{e}_1, \mathbf{e}_3, \mathbf{e}_2) +$
$a_2 b_1 c_3 A(\mathbf{e}_2, \mathbf{e}_1, \mathbf{e}_3) + a_2 b_3 c_1 A(\mathbf{e}_2, \mathbf{e}_3, \mathbf{e}_1) +$
$a_3 b_1 c_2 A(\mathbf{e}_3, \mathbf{e}_1, \mathbf{e}_2) + a_3 b_2 c_1 A(\mathbf{e}_3, \mathbf{e}_2, \mathbf{e}_1)$

$A(\mathbf{a}, \mathbf{b}, \mathbf{c}) = a_1 b_2 c_3 - a_1 b_3 c_2 +$
$+ a_2 b_1 c_3 - a_2 b_3 c_1 + a_3 b_1 c_2 - a_3 b_2 c_1$

$\begin{vmatrix} 1 & 2 & 3 \\ 2 & 3 & 4 \\ 3 & 4 & 6 \end{vmatrix} = 1 \cdot 3 \cdot 6 - 1 \cdot 4 \cdot 4 + 2 \cdot 2 \cdot 6$
$- 2 \cdot 4 \cdot 3 + 3 \cdot 2 \cdot 4 - 3 \cdot 3 \cdot 3 = -1$

On the next page, I will show you an easy to remember method for calculating determinants.

# How can you calculate a 3×3 determinant?

To remember the terms for the 3 × 3 determinant, the Sarrus rule is helpful.

$$\det(A) = \begin{vmatrix} a_{11} & a_{12} & a_{13} \\ a_{21} & a_{22} & a_{23} \\ a_{31} & a_{32} & a_{33} \end{vmatrix} =$$

$$a_{11}a_{22}a_{33} + a_{12}a_{23}a_{31} + a_{13}a_{21}a_{32}$$
$$-a_{31}a_{22}a_{13} - a_{32}a_{33}a_{11} - a_{33}a_{21}a_{12}$$

You write out the matrix in a grid and then repeat the first and second columns to the right. Then you add the three products of the diagonals from top left to bottom right (solid lines) and subtract the three products of the diagonals from bottom left to top right (dashed lines).

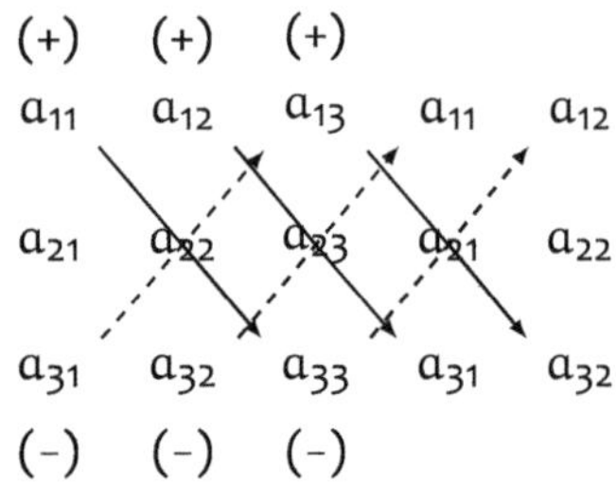


N.B. This rule only applies to 3 × 3 determinants; there is no similar rule for 4 × 4 determinants.

By factoring out you obtain the Laplace expansion for this determinant. You expand the determinant along the first column: you take $a_{11}$ times the minor that results when you delete the row and column of $a_{11}$, minus $a_{12}$ times the minor that results when you delete the row and column of $a_{12}$, etc. This works equally for any column or row and also for any size of the determinant. The signs are distributed in a checkerboard pattern.

$$\det(A) = +a_{11}(a_{22}a_{33} - a_{32}a_{23})$$
$$- a_{12}(a_{21}a_{33} - a_{31}a_{33})$$
$$+ a_{13}(a_{21}a_{32} - a_{31}a_{22})$$

$$\begin{vmatrix} + & - & + & \cdots \\ - & + & - & \cdots \\ \vdots & \vdots & \vdots & \ddots \end{vmatrix}$$

In this way, you can reduce an $n \times n$ determinant to $n$ determinants of size $(n - 1) \times (n - 1)$.

The 2 × 2 determinant also follows this pattern and you can remember it this way.

Often, you can simplify the calculation by adding or subtracting any multiple of a row (column) to another row (column).

$$\begin{vmatrix} 1 & 2 & 3 \\ 2 & 3 & 4 \\ 3 & 4 & 6 \end{vmatrix} = \xrightarrow[\;(3')=(3)-2\cdot(1)\;]{(2')=(2)-(1)} = \begin{vmatrix} 1 & 2 & 3 \\ 1 & 1 & 1 \\ 1 & 0 & 0 \end{vmatrix}$$

$$\xrightarrow[\text{along } 3^{\text{rd}} \text{ row}]{\text{Expanding}} = 1 \cdot \begin{vmatrix} 2 & 3 \\ 1 & 1 \end{vmatrix} = -1$$

# How can you use determinants to calculate areas and volumes?

The area of the parallelogram spanned by two vectors is the determinant of the two vectors.

$$A_{\square} = \det(\mathbf{a}, \mathbf{b}) = \begin{vmatrix} a_1 & b_1 \\ a_2 & b_2 \end{vmatrix}$$

From this, it follows that the area of the triangle with vertices $(0,0)$, $(a_1, a_2)$, and $(b_1, b_2)$ is half of that.

$$A_{\triangle OAB} = \frac{1}{2} \det(\mathbf{a}, \mathbf{b}) = \frac{1}{2} \begin{vmatrix} a_1 & b_1 \\ a_2 & b_2 \end{vmatrix}$$

If you represent the area of any triangle $ABC$ with $A(a_1, a_2)$, $B(b_1, b_2)$, $C(c_1, c_2)$ as the sum or difference of the areas of the triangles OBC, OBA, and OAC, the three $2 \times 2$ matrices add up according to the Laplace expansion theorem to a $3 \times 3$ matrix.

$$A_{\triangle ABC} = \frac{1}{2} \begin{vmatrix} a_1 & b_1 & c_1 \\ a_2 & b_2 & c_2 \\ 1 & 1 & 1 \end{vmatrix}$$

This follows from
$$A_{\triangle ABC} = A_{\triangle OBC} - A_{\triangle OAC} + A_{\triangle OAB}$$
$$= \tfrac{1}{2} \det(\mathbf{b}, \mathbf{c}) - \tfrac{1}{2} \det(\mathbf{a}, \mathbf{c})$$
$$+ \tfrac{1}{2} \det(\mathbf{a}, \mathbf{b}).$$

The two formulas mentioned above also apply equally in 3 dimensions, i.e. for a tetrahedron. All determinants have one more dimension and the prefactor is 1/6 instead of 1/2.

You can also determine the area of any triangle and the volume of any tetrahedron from the 3 or 6 edge lengths. If you are interested, look up "Heron's formula" or "Cayley-Menger determinant" on the internet.

Just as you can use the determinant $\det(\mathbf{a}, \mathbf{b})$ to check whether the vectors are linearly dependent, you can use the above formula for the area of a general triangle to determine whether 3 points are collinear.

$\det(\mathbf{a}, \mathbf{b}) = 0 \Leftrightarrow$ The vectors $\mathbf{a}, \mathbf{b}$ are linearly dependent.

$$\begin{vmatrix} a_1 & b_1 & c_1 \\ a_2 & b_2 & c_2 \\ 1 & 1 & 1 \end{vmatrix} = 0 \Leftrightarrow$$

The points $(a_1, a_2)$, $(b_1, b_2)$, $(c_1, c_2)$ are collinear.

This also works similarly for the linear dependence of 3 vectors with a $3 \times 3$ determinant, and for checking if four points are coplanar with a $4 \times 4$ determinant.

# What is the cross product? (I)

The cross product, also called vector product, is only defined in three dimensions and assigns to two vectors a third vector that is perpendicular to both and thus is the normal vector of the plane spanned by the two vectors.

The two conditions for orthogonality yield a linear system for the components of the cross product $x = a \times b$.

We solve the system and use a solution that avoids fractions. We directly use the vector on the right as the definition of the cross product, since its length has a simple interpretation.

If you take the square of the cross product and add the square of the scalar product, you get the product of the squares of the two vectors.

From this and with the trigonometric Pythagorean identity $\sin^2(\alpha) + \cos^2(\alpha) = 1$, you see that the length of the cross product is equal to the length of the first vector times the length of the second vector times the sine of the included angle.

$$a \times b \in \mathbb{R}^3$$
$$a, b \in \mathbb{R}^3$$
$$a \times b \perp a \Rightarrow (a \times b) \cdot a = 0$$
$$a \times b \perp b \Rightarrow (a \times b) \cdot b = 0$$
$$\Rightarrow (a \times b) \cdot (ra + sb) = 0 \text{ for all } r, s \in \mathbb{R}$$
$$\Rightarrow a \times b \text{ is perpendicular to the plane spanned by } a \text{ and } b.$$

$$a \cdot x = 0 \Rightarrow a_1 x_1 + a_2 x_2 + a_3 x_3 = 0 \quad (1)$$
$$b \cdot x = 0 \Rightarrow b_1 x_1 + b_2 x_2 + b_3 x_3 = 0 \quad (2)$$

$b_3 \cdot (2) - a_3 \cdot (1)$ eliminates $x_3$:
$$(a_1 b_3 - a_3 b_1) x_1 + (a_2 b_3 - a_3 b_2) x_2 = 0.$$

The solution is given by the "swapped" coefficients of $x_1$ and $x_2$ with a sign change:
$$x_1 = a_2 b_3 - a_3 b_2$$
$$x_2 = a_3 b_1 - a_1 b_3, \text{ and it follows}$$
$$x_3 = a_1 b_2 - a_2 b_1.$$

$$(a_2 b_3 - a_3 b_2)^2 + (a_3 b_1 - a_1 b_3)^2$$
$$+ (a_1 b_2 - a_2 b_1)^2 + (a_1 b_1 + a_2 b_2 + a_3 b_3)^2$$
$$= (a_1^2 + a_2^2 + a_3^2)(b_1^2 + b_2^2 + b_3^3)$$

You can easily verify it by expanding.

So $|a \times b|^2 + (a \cdot b)^2 = |a|^2 |b|^2$
$$\Rightarrow |a \times b|^2 = |a|^2 |b|^2 - (a \cdot b)^2$$

With $\varphi = \angle(a, b)$ it follows:
$$|a \times b|^2 = |a|^2 |b|^2 - |a|^2 |b|^2 \cos^2(\varphi)$$
$$= |a|^2 |b|^2 (1 - \cos^2(\varphi))$$
$$= |a|^2 |b|^2 \sin^2(\varphi)$$

$$\Rightarrow |a \times b| = |a| \, |b| \, |\sin(\varphi)|$$

The magnitude of the cross product is the area of the parallelogram spanned by $a$ and $b$.

# What is the cross product? (II)

You can remember the components of the cross product by using a formal determinant, where the $e_1$, $e_2$ and $e_3$ are the three unit vectors.

$$a \times b = \begin{pmatrix} a_2 b_3 - a_3 b_2 \\ a_3 b_1 - a_1 b_3 \\ a_1 b_2 - a_2 b_1 \end{pmatrix}$$

$$a \times b = \begin{vmatrix} e_1 & a_1 & b_1 \\ e_2 & a_2 & b_2 \\ e_3 & a_3 & b_3 \end{vmatrix}$$

Or you remember the first component $(a \times b)_1 = a_2 b_3 - a_3 b_2$. You get the second and third components by cyclic permutation of the indices.

By cyclic permutation
$$1 \to 2 \to 3 \to 1 \to \ldots$$
the $1^{st}$ component $(a_2 b_3 - a_3 b_2)$ becomes $(a_3 b_1 - a_1 b_3)$, then $(a_1 b_2 - a_2 b_1)$.

The direction of the cross product is defined so that $a$, $b$, and $a \times b$ in this order correspond to the thumb, index finger, and middle finger according to the right hand rule.

We say: $a$, $b$, and $a \times b$ form a right-handed system. The cross product is anticommutative:
$$a \times b = -b \times a \text{ and non-associative:}$$
$$a \times (b \times c) \neq (a \times b) \times c.$$

The scalar product of $a \times b$ with a third vector is called the scalar triple product. With the determinant representation of the cross product, you see that the scalar triple product is the determinant of the matrix whose columns are the three vectors. Therefore, it is also written as $[abc] = (a \times b) \cdot c$.

$$(a \times b) \cdot c = \begin{vmatrix} c_1 & a_1 & b_1 \\ c_2 & a_2 & b_2 \\ c_3 & a_3 & b_3 \end{vmatrix}$$

$$= \begin{vmatrix} a_1 & b_1 & c_1 \\ a_2 & b_2 & c_2 \\ a_3 & b_3 & c_3 \end{vmatrix} = [abc]$$

$[abc]$ is the volume of the parallelepiped spanned by the vectors $a$, $b$, and $c$. A parallelepiped is the 3-dimensional version of a parallelogram.

From this you see that the scalar triple product is invariant under cyclic permutation.

$$(a \times b) \cdot c = (b \times c) \cdot a = (c \times a) \cdot b = [abc]$$

There are some useful identities that can be verified in various ways.

$$a \times (b \times c) = b(a \cdot c) - c(a \cdot b)$$
$$(a \times b) \cdot (c \times d) = (a \cdot c)(b \cdot d) - (a \cdot d)(b \cdot c)$$

# What are the kernel and image of a linear map?

The kernel of a linear map $L : V \to W$ consists of all vectors $\mathbf{v} \in V$ with $L(\mathbf{v}) = \mathbf{0}$.

$\ker L = \{\mathbf{v} \in V\,|\, L(\mathbf{v}) = \mathbf{0}\}$

For $L : \mathbb{R}^3 \to \mathbb{R}^3$
$(v_1, v_2, v_3)^\top \mapsto (v_1, v_2, 0)^\top$:
$L(\mathbf{v}) = \mathbf{0} \Leftrightarrow v_1 = 0 \wedge v_2 = 0$

These vectors form a vector subspace of $V$.

$\ker L = \{(0, 0, v_3)^\top \,|\, v_3 \in \mathbb{R}\} \cong \mathbb{R} \subseteq \mathbb{R}^3$

A vector subspace $U \subseteq V$ is a non-empty subset of $V$ that is closed under addition and scalar multiplication.

The image of a linear map $L : V \to W$ consists of all vectors $w \in W$ for which there exists a $\mathbf{v} \in V$ with $\mathbf{w} = L(\mathbf{v})$.

$\operatorname{im} L = \{\mathbf{w} \in W\,|\, \mathbf{w} = L(\mathbf{v}), \mathbf{v} \in V\}$

For $L : \mathbb{R}^3 \to \mathbb{R}^3$:
$(v_1, v_2, v_3)^\top \mapsto (v_1, v_2, 0)^\top$

These vectors form a vector subspace of $W$.

$\operatorname{im} L = \{(v_1, v_2, 0)^\top \,|\, v_{1,2} \in \mathbb{R}\} \cong \mathbb{R}^2 \subseteq \mathbb{R}^3$

If a linear map $L : V \to W$ is injective, then by definition every element of $W$ has at most one preimage. Thus the kernel, as the preimage of the zero vector in $W$, consists only of the zero vector in $V$.

$L$ injective $\Rightarrow |\{\mathbf{v} \in V\,|\, L(\mathbf{v}) = \mathbf{w}\}| \leq 1$
$\Rightarrow |\ker L| = |\{\mathbf{v} \in V\,|\, L(\mathbf{v}) = \mathbf{0}\}| \leq 1$
Since $L(\mathbf{0}) = \mathbf{0}$, we have $\mathbf{0} \in \ker L$, and since $|\ker L| \leq 1$, there can be no other element in the kernel, so $\ker L = \{\mathbf{0}\}$.

Conversely, if the kernel consists only of the zero vector, then the linear map is injective.

For $L$ to be injective, it must hold that $L(\mathbf{v}) = L(\mathbf{w}) \Rightarrow \mathbf{v} = \mathbf{w}$.

This holds for $L$ with $\ker L = \{\mathbf{0}\}$ because
$L(\mathbf{v}) = L(\mathbf{w}) \Rightarrow L(\mathbf{v} - \mathbf{w}) = \mathbf{0}$
$\Rightarrow \mathbf{v} - \mathbf{w} = \mathbf{0},$
since $\ker L = \{\mathbf{0}\}$, and thus $\mathbf{v} = \mathbf{w}$.

For a general map, to prove injectivity, you must check for each element of the codomain that it has at most one preimage.

Thus, the kernel of a linear map $L$ consists only of the zero vector if and only if $L$ is injective.

$\ker L = \{\mathbf{0}\} \Leftrightarrow L$ is injective.

# What does the rank of a linear map or a matrix mean?

The rank of a linear map $L : \mathbb{R}^n \to \mathbb{R}^k$ is the dimension of the image of the linear map.

If the linear map is represented as a matrix $L \in \mathbb{R}^{k \times n}$, then the rank is equal to the dimension of the span of the columns, i.e. the number of linearly independent columns.

One can, for matrices over a field, show that this number is equal to the number of linearly independent rows.

You determine the rank by using the Gaussian elimination method to calculate the row echelon form. The number of nonzero rows is then the rank.

If and only if, for square matrices, the rank is less than the dimension, then the matrix is not invertible and its determinant is zero.

The rank is, in a sense, a refinement of $\det(M) = 0$. If the determinant is zero, the column vectors are linearly dependent. The rank is then less than the number of columns and indicates how many of the columns are linearly independent.

$\operatorname{rank} L = \dim(\operatorname{im} L)$
Sometimes it's written as $\operatorname{rk} L$.

We denote the matrix by $L$ and its columns by $c_i$, $i = 1, \ldots, n$.

$\operatorname{rank} L = \dim(\operatorname{span}(c_1, c_2, \ldots, c_n))$, since the column vectors are the images of the unit vectors and thus span the image.

From this and from the above, it follows:

$$\operatorname{rank} L \leq \min(n, k) \quad \text{for } L \in \mathbb{R}^{k \times n}.$$

With the Gaussian elimination method, you obtain a row echelon form from

$$\begin{pmatrix} 1 & 2 & 3 \\ 2 & 3 & 4 \\ 3 & 4 & 5 \end{pmatrix} \text{ to } \begin{pmatrix} 1 & 2 & 3 \\ 0 & -1 & -1 \\ 0 & -1 & -1 \end{pmatrix} \text{ to } \begin{pmatrix} 1 & 2 & 3 \\ 0 & -1 & -1 \\ 0 & 0 & 0 \end{pmatrix}.$$

Thus, the rank is equal to 2.

For $M \in \mathbb{R}^{n \times n}$ holds:
$\operatorname{rank} M < n$
$\Leftrightarrow \det(M) = 0$
$\Leftrightarrow M$ is not invertible.

$$\det \begin{pmatrix} 1 & 2 & 3 \\ 2 & 3 & 4 \\ 3 & 4 & 5 \end{pmatrix} = 0, \operatorname{rank} \begin{pmatrix} 1 & 2 & 3 \\ 2 & 3 & 4 \\ 3 & 4 & 5 \end{pmatrix} = 2$$

$$\det \begin{pmatrix} 1 & 2 & 3 \\ 1 & 2 & 3 \\ 1 & 2 & 3 \end{pmatrix} = 0, \operatorname{rank} \begin{pmatrix} 1 & 2 & 3 \\ 1 & 2 & 3 \\ 1 & 2 & 3 \end{pmatrix} = 1$$

$$\det \begin{pmatrix} 0 & 0 & 0 \\ 0 & 0 & 0 \\ 0 & 0 & 0 \end{pmatrix} = 0, \operatorname{rank} \begin{pmatrix} 0 & 0 & 0 \\ 0 & 0 & 0 \\ 0 & 0 & 0 \end{pmatrix} = 0$$

# What does the dimension theorem state?

The dimension theorem states that for a linear map $L : V \to W$, the dimension of the kernel plus the dimension of the image is equal to the dimension of $V$.

$$\dim \operatorname{Ker} L + \dim \operatorname{im} L = \dim V$$

For $L : \mathbb{R}^3 \to \mathbb{R}^2$
$$(v_1, v_2, v_3)^\top \mapsto (v_1, v_2)^\top :$$

$$\underbrace{\dim \operatorname{Ker} L}_{\dim \mathbb{R}^1 = 1} + \underbrace{\dim \operatorname{im} L}_{\dim \mathbb{R}^2 = 2} = \underbrace{\dim V}_{\dim \mathbb{R}^3 = 3}$$

You can understand it as follows: If the kernel has dimension $m \leq n = \dim V$, you can choose a basis $\{b_1, b_2, \ldots, b_m\}$ of the kernel. There are then $m - n$ further basis vectors $\{b_{m-n+1}, b_2, \ldots, b_n\}$, and the image is spanned by $\{L(b_{m-n+1}), \ldots, L(b_n)\}$. Since these vectors are linearly independent, the dimension is $\dim \operatorname{im} L = n - m = \dim V - \dim \ker L$.

$$V \left\{ \begin{array}{l} \ker L \left\{ \begin{array}{ll} b_1 \mapsto & L(b_1) = 0 \\ \vdots & \vdots \\ b_m \mapsto & L(b_m) = 0 \end{array} \right. \\ \left. \begin{array}{ll} b_{m+1} \mapsto & L(b_{m+1}) \\ \vdots & \vdots \\ b_n \mapsto & L(b_n) \end{array} \right\} \operatorname{im} L \end{array} \right.$$

Since the dimension of the image of $L$ is equal to the rank of $L$, this is also called the rank-nullity theorem: $\dim \operatorname{im} L = \operatorname{rank} L$, so

$$\dim \ker L + \dim \operatorname{rank} L = \dim V.$$

If a linear map is injective, i.e. $\dim \ker L = 0$, then $\dim \operatorname{im} L = \dim V$.

$$V \xrightarrow{\;\;L\;\;} W$$

L is injective
$\dim W \geq \dim V$

If a linear map is surjective, i.e. $\dim \operatorname{im} L = \dim W$, then we have $\dim \operatorname{im} L = 0$, and therefore $\dim \ker L = \dim V - \dim W$.

$$V \xrightarrow{\;\;L\;\;} W$$

L is surjective
$\dim V \geq \dim W$

If a linear map is bijective, then we have $\dim \ker L = 0$, and therefore $\dim \operatorname{im} L = \dim V = \dim W$.

$$V \xrightarrow{\;\;L\;\;} W$$

L is bijective
$\dim W = \dim V$

If a linear map has a maximal kernel, i.e. $\dim \ker L = \dim V$, then $\dim \operatorname{im} L = 0$.

$$V \xrightarrow{\;\;L\;\;} 0$$

L is the zero map
$\dim \ker L = \dim V$

# What does the dimension theorem say for endomorphisms?

An endomorphism is a linear map from a vector space $V$ to itself.

For endomorphisms of finite-dimensional vector spaces, the dimension of the kernel is therefore equal to the dimension of the vector space minus the dimension of the image.

If the kernel is the null space, then the image is the entire vector space. Therefore, for linear maps $V \to V$ ($\dim V < \infty$), injectivity and surjectivity are equivalent.

A rotation $\mathbb{R}^2 \to \mathbb{R}^2$ has a trivial kernel, so it's injective and thus also surjective, i.e. bijective.

An injective map between finite-dimensional vector spaces is therefore automatically also surjective and thus bijective. Similarly, a surjective map is also injective and thus bijective.

For infinite-dimensional vector spaces, this does not hold.

Examples of endomorphisms of $\mathbb{R}^2$ are rotations and projections.

$$\dim \ker L = \dim V - \dim \operatorname{im} L$$

For the projection
$$P : \mathbb{R}^2 \to \mathbb{R}^2, (x, y)^\top \mapsto (x, 0)^\top$$
we have $\dim \operatorname{im} P = 1$ and thus $\dim \ker L = \dim \mathbb{R}^2 - \dim \operatorname{im} P = 2 - 1 = 1$, the kernel is the 1-dimensional y-axis.

For $L : V \to V$, $\dim V < \infty$:
injective $\Leftrightarrow$ surjective

This is the same as the fact that for maps between finite sets $f : A \to A$, injectivity implies surjectivity and vice versa. For infinite sets, this does not hold.

The right shift on the vector space of real sequences $S_r : V \to V$
$$S_r : (a_1, a_2, a_3, \ldots) \mapsto (0, a_1, a_2, \ldots)$$
is injective but not surjective.

The left shift
$$S_l : (a_1, a_2, a_3, \ldots) \mapsto (a_2, a_3, a_4, \ldots)$$
is surjective but not injective.

# How to calculate the solution of a $3 \times 3$ system?
## I: Example with a unique solution

To solve a system of linear equations with three equations and three variables, you leave the 1st equation as it is.

$$\begin{array}{rcll} x_1 + 2x_2 + 3x_3 &=& 6 & (1) \\ 2x_1 + 3x_2 + 4x_3 &=& 7 & (2) \\ 3x_1 + 4x_2 + 6x_3 &=& 8 & (3) \end{array}$$

From the 2nd equation, you subtract a suitable multiple $(2\times)$ of the 1st equation, such that the coefficient of $x_1$ disappears, giving a new equation in the 2nd row, $(2')$.

$$\begin{array}{rcll} x_1 + 2x_2 + 3x_3 &=& 6 & \\ -x_2 - 2x_3 &=& -5 & (2) - 2 \cdot (1) \\ 3x_1 + 4x_2 + 6x_3 &=& 8 & \end{array}$$

You do the same with the 3rd equation and obtain a new equation in the 3rd row, $(3')$, which eliminates $x_1$ from the 3rd row. Therefore from the 2nd row on, only two variables remain.

$$\begin{array}{rcll} x_1 + 2x_2 + 3x_3 &=& 6 & \\ -x_2 - 2x_3 &=& -5 & (2') \\ -2x_2 - 3x_3 &=& -10 & (3) - 3 \cdot (1) \end{array}$$

Finally, you subtract a suitable multiple $(2\times)$ of the new 2nd equation from the 3rd equation, so that in the 3rd equation only one variable remains.

$$\begin{array}{rcll} x_1 + 2x_2 + 3x_3 &=& 6 & \\ -x_2 - 2x_3 &=& -5 & \\ x_3 &=& 0 & (3') - 2 \cdot (2') \end{array}$$

This is the row echelon form.

You can then directly solve the 3rd equation for $x_3$, then substitute into the 2nd equation to get $x_2$, and then substitute both into the 1st equation to get $x_1$.

Calculation of the solution:
from $(3'') \Rightarrow x_3 = 0$
in $(2') \Rightarrow x_2 = 5$
in $(1) \Rightarrow x_1 = -4$
The solution is $(-4, 5, 0)^\top$.

This is an example where there is a unique solution. The homogeneous equation (when the right-hand side is all zeros) also has a unique solution, the trivial solution $x_1 = x_2 = x_3 = 0$, which always exists, but in this case is the only one.

Here, the rank of the matrix is 3, so there are $3 - 3 = 0$, i.e. no free parameters, so the homogeneous system has only the trivial solution, and the inhomogeneous system has a unique solution.

# How to calculate the solution of a $3\times3$ system? II: Example with a 1-dimensional solution space

We consider a very similar linear system of equations with three equations and three variables.

You proceed exactly as in the example on the previous page and eliminate $x_1$ from the $2^{nd}$ equation.

You do the same with the third equation, which eliminates $x_1$ from the $3^{rd}$ row. Thus from the $2^{nd}$ row on, only two variables remain.

Now, as before, you subtract $2\times$ the new $2^{nd}$ equation from the $3^{rd}$ equation to have only one variable left in the $3^{rd}$ equation. However, in this example this variable also disappears, leaving a zero row.

Thus, you can choose $x_3$ as a free parameter, for example $x_3 = s \in \mathbb{R}$, and substitute it into the $2^{nd}$ equation, then substitute both into the $1^{st}$ equation to obtain $x_1$.

This is an example with a 1-dimensional solution space. The homogeneous equation (when the right-hand side is all zeros) has a 1-dimensional vector space as its solution: $x_1 = s, x_2 = -2s, x_3 = s$, i.e. $\mathbf{x} = (1, -2, 1)^{\top} s$, $s \in \mathbb{R}$, that is, a line through the origin.

$$
\begin{array}{rcll}
x_1 + 2x_2 + 3x_3 &=& 6 & (1)\\
2x_1 + 3x_2 + 4x_3 &=& 7 & (2)\\
3x_1 + 4x_2 + 5x_3 &=& 8 & (3)
\end{array}
$$

$$
\begin{array}{rcll}
x_1 + 2x_2 + 3x_3 &=& 6 & \\
-x_2 - 2x_3 &=& -5 & (2) - 2\cdot(1)\\
3x_1 + 4x_2 + 5x_3 &=& 8 &
\end{array}
$$

$$
\begin{array}{rcll}
x_1 + 2x_2 + 3x_3 &=& 6 & \\
-x_2 - 2x_3 &=& -5 & (2')\\
-2x_2 - 4x_3 &=& -10 & (3) - 3\cdot(1)
\end{array}
$$

$$
\begin{array}{rcll}
x_1 + 2x_2 + 3x_3 &=& 6 & \\
-x_2 - 2x_3 &=& -5 & \\
0 &=& 0 & (3') - 2\cdot(2')
\end{array}
$$

This is the row echelon form.

Calculation of the solution:
from $(3'') \Rightarrow x_3 = s \in \mathbb{R}$
in $(2') \Rightarrow x_2 = -2s + 5$
in $(1) \Rightarrow x_1 = s - 4$
The solution set is a line.

Here, the rank of the matrix is 2, so there are $3 - 2 = 1$ free parameters. The homogeneous system has a 1-dimensional vector space as its solution, and the inhomogeneous system also has a line as its solution.

# How to calculate the solution of a $3 \times 3$ system? III: Example with empty solution set

We again consider a system of linear equations with three equations and three variables. The coefficients are as in the previous example, but in the last equation the right-hand side is 9 instead of 8.

$$\begin{array}{rcll} x_1 + 2x_2 + 3x_3 &=& 6 & (1) \\ 2x_1 + 3x_2 + 4x_3 &=& 7 & (2) \\ 3x_1 + 4x_2 + 5x_3 &=& 9 & (3) \end{array}$$

$$\begin{array}{rcll} x_1 + 2x_2 + 3x_3 &=& 6 & \\ -x_2 - 2x_3 &=& -5 & (2) - 2 \cdot (1) \\ 3x_1 + 4x_2 + 5x_3 &=& 9 & \end{array}$$

You proceed exactly as in the previous example and eliminate $x_1$ from the 2nd equation.

You do the same with the 3rd equation, and again from the 2nd row on, only two variables remain.

$$\begin{array}{rcll} x_1 + 2x_2 + 3x_3 &=& 6 & \\ -x_2 - 2x_3 &=& -5 & (2') \\ -2x_2 - 4x_3 &=& -9 & (3) - 3 \cdot (1) \end{array}$$

Now, as before, you subtract twice the new 2nd equation from the 3rd equation to have only one variable left in the 3rd equation. However, in this example this variable also disappears, while on the right-hand side a 1 remains.

$$\begin{array}{rcll} x_1 + 2x_2 + 3x_3 &=& 6 & \\ -x_2 - 2x_3 &=& -5 & \\ 0 &=& 1 & (3') - 2 \cdot (2') \end{array}$$

This is the row echelon form.

Thus, the last equation is false regardless of the variables, and the system has no solution.

The solution set is the empty set.

This is therefore an example with an empty solution set. The homogeneous equation (when the right-hand side is all zeros) is as in the previous example and therefore has the same one-dimensional solution set: $x_1 = s$, $x_2 = -2s$, $x_3 = s$.

Here, the rank of the matrix is 2, so there are $3 - 2 = 1$ free parameters. The homogeneous system has a one-dimensional vector space as its solution, but the inhomogeneous system has no solution.

# How do you represent a system of linear equations as a matrix?

A linear system of equations is simply a linear matrix equation $A\mathbf{x} = \mathbf{b}$.

The matrix $A$ is the matrix of coefficients. Each row corresponds to the left side of an equation, each column to an unknown.

The vector $\mathbf{b}$ represents the right hand side of the system of equations. If $\mathbf{b} = \mathbf{0}$, you have a homogeneous system of linear equations; otherwise, it is inhomogeneous.

You can solve a system of linear equations directly in this matrix notation, which saves you from writing out the unknowns each time.

Here is the example from above with the unique solution. If individual rows of the coefficient matrix have leading zeros, you do not write them out. This way, you can directly recognize the row echelon form.

For $A \in \mathbb{R}^{3\times3}$, $\mathbf{b}, \mathbf{x} \in \mathbb{R}^3$, the vector equation $A\mathbf{x} = \mathbf{b}$ corresponds to the following 3 equations:

$$a_{11}x_1 + a_{12}x_2 + a_{13}x_3 = b_1$$
$$a_{21}x_1 + a_{22}x_2 + a_{23}x_3 = b_2$$
$$a_{31}x_1 + a_{32}x_2 + a_{33}x_3 = b_3$$

$$\left(\begin{array}{ccc|c} 1 & 2 & 3 & 6 \\ 2 & 3 & 4 & 7 \\ 3 & 4 & 6 & 8 \end{array}\right) \begin{array}{l} (1) \\ (2) \\ (3) \end{array}$$

$$\left(\begin{array}{ccc|c} 1 & 2 & 3 & 6 \\  & -1 & -2 & -5 \\ 3 & 4 & 6 & 8 \end{array}\right) \begin{array}{l} (2) - 2 \cdot (1) \end{array}$$

$$\left(\begin{array}{ccc|c} 1 & 2 & 3 & 6 \\  & -1 & -2 & -5 \\  & -2 & -3 & -10 \end{array}\right) \begin{array}{l} (2') \\ (3) - 3 \cdot (1) \end{array}$$

$$\left(\begin{array}{ccc|c} 1 & 2 & 3 & 6 \\  & -1 & -2 & -5 \\  &  & 1 & 0 \end{array}\right) \begin{array}{l} (3') - 2 \cdot (2') \end{array}$$

# What is the structure of the set of solutions of a linear system?

We distinguish between homogeneous and inhomogeneous systems.

For a homogeneous system of linear equations, there are only zeros on the right hand side.

homogeneous system: $Ax = 0$
In vector notation, the right hand side is the zero vector.
It always has at least the solution $x = 0$.

You can easily see that the solutions of a homogeneous system of linear equations form a vector space.

If $x^{(1)}$ and $x^{(2)}$ are solutions, then $Ax^{(1)} = 0$ and $Ax^{(2)} = 0$, therefore:

$$A(x^{(1)} + x^{(2)}) = Ax^{(1)} + Ax^{(2)} = 0 \text{ and}$$
$$A(rx^{(1)}) = rAx^{(1)} = 0.$$
Thus the solutions form a vector space.

For an inhomogeneous system of linear equations, the right hand side consists of arbitrary numbers, at least one of which is not zero.

inhomogeneous system:
$Ax = b,\ b \neq 0$
In vector notation, the right hand side is any vector not equal to the zero vector.

The general solution of an inhomogeneous system is a particular solution of the inhomogeneous system plus the general solution of the homogeneous system.

$$x_{\text{inh, gen}} = x_{\text{inh, part}} + x_{\text{hom, gen}}$$

You can see that every solution must be of this form, if you represent a general solution as the sum of a particular solution and a "remaining solution". You can see that the remaining solution must solve the homogeneous system.

$$x_{\text{inh, gen}} = x_{\text{inh, part}} + x_{\text{rem}}$$
$$Ax_{\text{inh, gen}} = A(x_{\text{inh, part}} + x_{\text{rem}})$$
$$= \underbrace{Ax_{\text{inh, part}}}_{b} + Ax_{\text{rem}} = b$$

$\Rightarrow Ax_{\text{rem}} = b - b = 0$
$x_{\text{rem}}$ solves the homogeneous system.

Since linear differential equations can be represented as $Ly(x) = f(x)$ and $L$ is linear, a similar statement holds for the solutions.

$$y_{\text{inh, gen}} = y_{\text{inh, part}} + y_{\text{hom, gen}}$$

# How do you calculate the inverse of a $3 \times 3$ matrix?

Calculating the inverse of a $3 \times 3$ matrix is equivalent to solving three systems of linear equations, where the inhomogeneities are the three unit vectors, respectively.

On the right, you see the calculation of the inverse of the coefficient matrix from the previous example.

You perform the row operations as above, aiming to get the identity matrix on the left.

After the last step, at the bottom, you can read off the inverse matrix on the right.

To make sure you haven't made a calculation error, you can multiply the two matrices together: the result must be the identity matrix.

Such "checks", where you verify whether the original requirements are satisfied, are useful in many calculations, for example: After plugging in the found solution, are the original equations of a system of linear equations true? Does the point through which a plane is supposed to pass satisfy the plane equation you found?

$$\left( \begin{array}{ccc|ccc} 1 & 2 & 3 & 1 & 0 & 0 \\ 2 & 3 & 4 & 0 & 1 & 0 \\ 3 & 4 & 6 & 0 & 0 & 1 \end{array} \right) \begin{array}{l} (1) \\ (2) \\ (3) \end{array}$$

$$\left( \begin{array}{ccc|ccc} 1 & 2 & 3 & 1 & 0 & 0 \\ 0 & -1 & -2 & -2 & 1 & 0 \\ 3 & 4 & 6 & 0 & 0 & 1 \end{array} \right) \begin{array}{l} \\ (2) - 2 \cdot (1) \\ \end{array}$$

$$\left( \begin{array}{ccc|ccc} 1 & 2 & 3 & 1 & 0 & 0 \\ 0 & -1 & -2 & -2 & 1 & 0 \\ 0 & -2 & -3 & -3 & 0 & 1 \end{array} \right) \begin{array}{l} \\ (2') \\ (3) - 3 \cdot (1) \end{array}$$

$$\left( \begin{array}{ccc|ccc} 1 & 2 & 3 & 1 & 0 & 0 \\ 0 & -1 & -2 & -2 & 1 & 0 \\ 0 & 0 & 1 & 1 & -2 & 1 \end{array} \right) \begin{array}{l} \\ \\ (3') - 2 \cdot (2') \end{array}$$

$$\left( \begin{array}{ccc|ccc} 1 & 2 & 3 & 1 & 0 & 0 \\ 0 & -1 & 0 & 0 & -3 & 2 \\ 0 & 0 & 1 & 1 & -2 & 1 \end{array} \right) \begin{array}{l} \\ (2') + 2 \cdot (3'') \\ \end{array}$$

$$\left( \begin{array}{ccc|ccc} 1 & 0 & 0 & -2 & 0 & 1 \\ 0 & 1 & 0 & 0 & 3 & -2 \\ 0 & 0 & 1 & 1 & -2 & 1 \end{array} \right) \begin{array}{l} \\ -1 \cdot (2'') \\ \end{array}$$

# How to bring a linear system into row echelon form?

The procedure shown in the examples above also works in general for $m$ linear equations in $n$ variables.

You choose as the first equation an equation whose first coefficient is not zero.

If there isn't one, you renumber the variables (i.e. swap the columns). If the first equation consists only of zeros, move it to the bottom.

You leave the first equation as is. From the second equation, you subtract such a multiple of the first equation that the coefficient of the first variable becomes zero. You continue in this way.

You obtain the row echelon form, in which each row starts with more zeros than the row above.

The number of dark boxes is equal to the number $r$ of nonzero rows, and equal to the rank of the matrix.

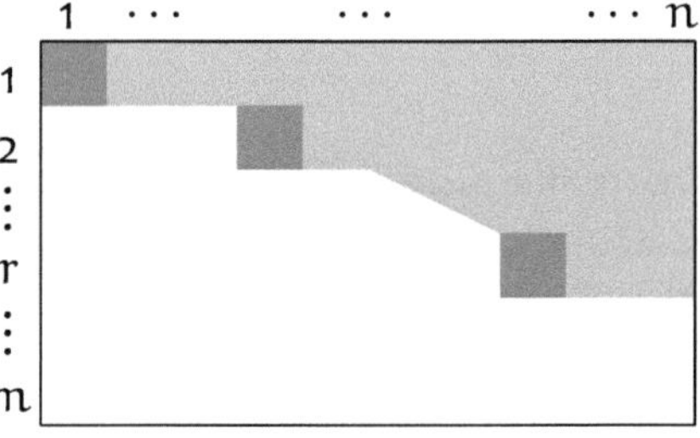

If there is "only" one more zero in each row, the horizontal dots have length zero and the dark boxes form a $45°$ diagonal. This is the case when the rank equals the number of variables and the homogeneous system has only the trivial solution $(0, 0, \ldots, 0)^\top$.

$n$:   number of variables
$m$:   number of rows
$r$:   rank

dark gray: numbers $\neq 0$
light gray:     arbitrary numbers
white:     zeros

number of columns
    with dark boxes: $r$
    without dark boxes: $n - r$

The $n - r$ unknowns corresponding to the columns without a dark box can be chosen as free parameters.

The solution space of the homogeneous system is an $(n-r)$-dimensional subspace.

# What does the rank of the augmented coefficient matrix say about the solution?

We consider the system $A\mathbf{x} = \mathbf{b}$ $\qquad A \in \mathbb{R}^{n \times n}, \ \mathbf{x}, \mathbf{b} \in \mathbb{R}^n$

**$\mathbf{b} = \mathbf{0}$: homogeneous system**

For $\det(A) \neq 0$, i.e. $rk(A) = n$:
$\mathbf{x} = \mathbf{0}$ is the unique solution.

$$\begin{pmatrix} 1 & 2 \\ 2 & 3 \end{pmatrix} \begin{pmatrix} x_1 \\ x_2 \end{pmatrix} = \begin{pmatrix} 0 \\ 0 \end{pmatrix} \Rightarrow \mathbf{x} = \begin{pmatrix} 0 \\ 0 \end{pmatrix}$$

For $\det(A) = 0$, i.e. $rk(A) < n$:
You can choose $n - rk(A) > 0$ free parameters. Thus the solution space is an $(n - rk(A))$-dimensional vector space.

$$\begin{pmatrix} 1 & 2 \\ 2 & 4 \end{pmatrix} \begin{pmatrix} x_1 \\ x_2 \end{pmatrix} = \begin{pmatrix} 0 \\ 0 \end{pmatrix}$$

$$\Rightarrow \mathbf{x} = r \begin{pmatrix} 2 \\ -1 \end{pmatrix}$$

1-dimensional solution space

**$\mathbf{b} \neq \mathbf{0}$: inhomogeneous system**

For $\det(A) \neq 0$, so $rk(A) = n$:
$\mathbf{x} = A^{-1}\mathbf{b}$ is the unique solution.

$$\begin{pmatrix} 1 & 2 \\ 2 & 3 \end{pmatrix} \begin{pmatrix} x_1 \\ x_2 \end{pmatrix} = \begin{pmatrix} 1 \\ 1 \end{pmatrix}$$

$$\Rightarrow \mathbf{x} = \begin{pmatrix} -3 & 2 \\ 2 & -1 \end{pmatrix} \begin{pmatrix} 1 \\ 1 \end{pmatrix} = \begin{pmatrix} -1 \\ 1 \end{pmatrix}$$

For $\det(A) = 0$, so $rk(A) < n$:

a) If $rk(A) = rk(A|\mathbf{b})$:
The solution space is $(n - rk(A))$-dimensional: The general solution of the inhomogeneous system = particular solution of the inhomogeneous system + a general solution of the homogeneous system.

$$\begin{pmatrix} 1 & 2 \\ 2 & 4 \end{pmatrix} \begin{pmatrix} x_1 \\ x_2 \end{pmatrix} = \begin{pmatrix} 3 \\ 6 \end{pmatrix}$$

$\mathbf{b}$ is linearly dependent on the other two columns
$\Rightarrow rk(A) = rk(A|\mathbf{b}) = 1$
$$\Rightarrow \mathbf{x} = \begin{pmatrix} 1 \\ 1 \end{pmatrix} + r \begin{pmatrix} 2 \\ -1 \end{pmatrix}$$
1-dimens. solution space

b) If $rk(A) \neq rk(A|\mathbf{b})$:
There is no solution.

$$\begin{pmatrix} 1 & 2 \\ 2 & 4 \end{pmatrix} \begin{pmatrix} x_1 \\ x_2 \end{pmatrix} = \begin{pmatrix} 3 \\ 4 \end{pmatrix}$$

$\mathbf{b}$ is linearly independent of the other two columns
$\Rightarrow rk(A|\mathbf{b}) = 2 > rk(A) = 1$
$\Rightarrow$ There is no solution.

# What does the dimension theorem say specifically for $A\mathbf{x} = 0$ and $\mathbf{x} \in \mathbb{R}^3$?

The solution space of the homogeneous system $A\mathbf{x} = \mathbf{0}$ is the kernel of the linear map $A$.

This is exactly how the kernel is defined.

The dimension theorem states that the dimension of the solution space is equal to the dimension of the whole space $\mathbb{R}^3$ (from which the $\mathbf{x}$s come), minus the rank of the matrix $A$.

$$\dim \ker A = \dim V - \dim \operatorname{im} A$$
$$\phantom{\dim \ker A} = \quad 3 - \operatorname{rk} A$$

| Rank of matrix $A$ | Dimension of the solution space | Example for matrix $A$ | Solution space for the example |
|---|---|---|---|
| 0 | $3 - 0 = 3$ | $\begin{pmatrix} 0 & 0 & 0 \\ 0 & 0 & 0 \\ 0 & 0 & 0 \end{pmatrix}$ | $x_1, x_2, x_3 \in \mathbb{R}, 0 = 0$ <br> entire space $\mathbb{R}^3$ |
| 1 | $3 - 1 = 2$ | $\begin{pmatrix} 0 & 0 & 0 \\ 0 & 0 & 0 \\ 0 & 0 & 1 \end{pmatrix}$ | $x_1, x_2 \in \mathbb{R}, x_3 = 0$ <br> plane |
| 2 | $3 - 2 = 1$ | $\begin{pmatrix} 0 & 0 & 0 \\ 0 & 1 & 0 \\ 0 & 0 & 1 \end{pmatrix}$ | $x_1 \in \mathbb{R}, x_2 = x_3 = 0$ <br> line |
| 3 | $3 - 3 = 0$ | $\begin{pmatrix} 1 & 0 & 0 \\ 0 & 1 & 0 \\ 0 & 0 & 1 \end{pmatrix}$ | $x_1 = x_2 = x_3 = 0$ <br> point |

Thus, each equation reduces the solution space from full dimension 3 by one dimension.

Each equation corresponds to a nonzero row of $A$.

This holds only if the equations are linearly independent of each other, which is exactly what the rank measures.

It's clear that, for example, two identical equations only reduce the dimension by one.

# How do you calculate the intersection point of two lines?

In the plane, two lines have an intersection point unless they are parallel.

If both lines are given in parametric form, set the two equations equal component-wise to obtain two equations for the two unknown parameters. Solve the resulting system of linear equations and substitute one parameter into the corresponding parametric form and the other into the other as a check.

$$g: \mathbf{x} = \begin{pmatrix} 0 \\ 1 \end{pmatrix} + r \begin{pmatrix} 1 \\ -1 \end{pmatrix}$$

$$h: \mathbf{x} = \begin{pmatrix} 0 \\ -1 \end{pmatrix} + s \begin{pmatrix} 1 \\ 2 \end{pmatrix}$$

$$\mathbf{x}_S = g \cap h: r \begin{pmatrix} 1 \\ -1 \end{pmatrix} = \begin{pmatrix} 0 \\ -2 \end{pmatrix} + s \begin{pmatrix} 1 \\ 2 \end{pmatrix}$$

$$\Rightarrow r = 0 + s, \; -r = -2 + 2s$$
$$\Rightarrow r = s, \; -s = -2 + 2s$$
$$\Rightarrow s = 2/3, \; r = 2/3$$

$$\Rightarrow \mathbf{x}_S = \begin{pmatrix} 0 \\ 1 \end{pmatrix} + \frac{2}{3} \begin{pmatrix} 1 \\ -1 \end{pmatrix} = \begin{pmatrix} 2/3 \\ 1/3 \end{pmatrix}$$

$r$ in $g$ substituted yields the same point.

If you have a line in HNF (or in explicit form) and the second line in parametric form, substitute the parametric form into the HNF to obtain a linear equation for the parameter. Solve for the parameter and substitute it into the parametric form.

$$g: \frac{1}{\sqrt{2}} \begin{pmatrix} 1 \\ 1 \end{pmatrix} \cdot \mathbf{x} - \frac{1}{\sqrt{2}} = 0$$

$$h: \mathbf{x} = \begin{pmatrix} 0 \\ -1 \end{pmatrix} + s \begin{pmatrix} 1 \\ 2 \end{pmatrix}$$

$$\frac{1}{\sqrt{2}} \begin{pmatrix} 1 \\ 1 \end{pmatrix} \cdot \left( \begin{pmatrix} 0 \\ -1 \end{pmatrix} + s \begin{pmatrix} 1 \\ 2 \end{pmatrix} \right) - \frac{1}{\sqrt{2}} = 0 \mid \cdot \sqrt{2}$$

$$-1 + 3s - 1 = 0 \Rightarrow s = \tfrac{2}{3} \Rightarrow \mathbf{x}_S = \begin{pmatrix} 2/3 \\ 1/3 \end{pmatrix}$$

In 3 dimensions, you can proceed similarly.

If the three equations for the two unknowns have no solution, the two lines are parallel or skew.

You can distinguish these two cases by the direction vectors: if the two direction vectors are parallel, the lines are parallel; otherwise, they are skew.

# How do you calculate the point of line-plane-intersections?

If you are given the line and the plane in parametric form, set the two equations equal component-wise to obtain three linear equations for the three unknown parameters. Solve the system of linear equations and, if there is a unique solution, substitute the parameters into the corresponding parametric form to obtain the intersection point.

$$P: \mathbf{x} = \begin{pmatrix} 0 \\ 0 \\ 1 \end{pmatrix} + r \begin{pmatrix} 1 \\ 0 \\ -1 \end{pmatrix} + s \begin{pmatrix} 0 \\ 1 \\ -1 \end{pmatrix}$$

$$g: \mathbf{x} = \begin{pmatrix} 1 \\ 1 \\ 1 \end{pmatrix} + t \begin{pmatrix} 1 \\ 1 \\ 1 \end{pmatrix}$$

It's important that you name the three parameters differently.

$\mathbf{x} = P \cap g:$

$r = 1 + t, \ s = 1 + t, \ -r - s = t$

$\Rightarrow r = s = 1 + t, \ -2(1 + t) = t$

$\Rightarrow t = -\frac{2}{3}, r = s = \frac{1}{3}$

$$\Rightarrow \mathbf{x} = \begin{pmatrix} 1/3 \\ 1/3 \\ 1/3 \end{pmatrix} \text{ is the intersection.}$$

If you are given the plane in HNF (or in explicit form) and the line in parametric form, substitute the parametric form into the HNF and solve the equation for the parameter. Then proceed as above.

$$P: \frac{1}{\sqrt{3}} \begin{pmatrix} 1 \\ 1 \\ 1 \end{pmatrix} \cdot \mathbf{x} - \frac{1}{\sqrt{3}} = 0$$

$$g: \mathbf{x} = \begin{pmatrix} 1 \\ 1 \\ 1 \end{pmatrix} + r \begin{pmatrix} 1 \\ 1 \\ 1 \end{pmatrix}$$

$$\frac{1}{\sqrt{3}} \begin{pmatrix} 1 \\ 1 \\ 1 \end{pmatrix} \cdot \left( \begin{pmatrix} 1 \\ 1 \\ 1 \end{pmatrix} + r \begin{pmatrix} 1 \\ 1 \\ 1 \end{pmatrix} \right) - \frac{1}{\sqrt{3}} = 0 \Big| \cdot \sqrt{3}$$

$$3 + 3r - 1 = 0 \Rightarrow r = -\frac{2}{3} \Rightarrow \mathbf{x} = \begin{pmatrix} 1/3 \\ 1/3 \\ 1/3 \end{pmatrix}$$

If there is no solution, the line is parallel to the plane.

$$h: \mathbf{x} = \begin{pmatrix} 1 \\ 1 \\ 1 \end{pmatrix} + r \begin{pmatrix} 1 \\ 0 \\ -1 \end{pmatrix} \text{ and } P \text{ as above}$$

$\Rightarrow 3 - 1 = 0,$ so contradiction $\Rightarrow h \| P.$

# How do you calculate the line of intersection between planes?

If you are given both planes in parametric form, set the two equations equal componentwise and to obtain three linear equations for the four unknown parameters. Thus in the generic case one parameter remains, by which you can express the solution $\mathbf{x}$; this is the line of intersection in parametric form.

$$\mathbf{x} = \begin{pmatrix} 1 \\ 0 \\ 0 \end{pmatrix} + r \begin{pmatrix} 1 \\ -1 \\ 0 \end{pmatrix} + s \begin{pmatrix} 1 \\ 0 \\ -1 \end{pmatrix}$$

$$\mathbf{x} = \begin{pmatrix} 1 \\ 0 \\ 0 \end{pmatrix} + t \begin{pmatrix} 1 \\ -1 \\ 0 \end{pmatrix} + u \begin{pmatrix} 0 \\ 0 \\ 1 \end{pmatrix}$$

$$\Rightarrow 1 + r + s = 1 + t, \; -r = -t, \; -s = u$$
$$\Rightarrow 1 - t + u = 1 + t \Rightarrow u = 0, t \in \mathbb{R}$$

$$\Rightarrow \mathbf{x} = \begin{pmatrix} 1 \\ 0 \\ 0 \end{pmatrix} + t \begin{pmatrix} 1 \\ -1 \\ 0 \end{pmatrix}, \; t \in \mathbb{R}$$

If you are given one plane in HNF and one in parametric form, then you substitute the parametric form into the HNF to obtain an equation with two unknown parameters. Thus you can eliminate one parameter. Substituting the parameters into the parametric form gives you the line of intersection.

$$\frac{1}{\sqrt{3}} \begin{pmatrix} 1 \\ 1 \\ 1 \end{pmatrix} - \frac{1}{\sqrt{3}} = 0$$

$$\mathbf{x} = \begin{pmatrix} 1 \\ 0 \\ 0 \end{pmatrix} \cdot \mathbf{x} + r \begin{pmatrix} 1 \\ -1 \\ 0 \end{pmatrix} + s \begin{pmatrix} 0 \\ 0 \\ 1 \end{pmatrix}$$

$$\Rightarrow (1+r+0)+(0+-r+0)+(0+0+s) = 1$$
$$\Rightarrow s = 0, r \in \mathbb{R}$$

$$\Rightarrow \mathbf{x} = \begin{pmatrix} 1 \\ 0 \\ 0 \end{pmatrix} + r \begin{pmatrix} 1 \\ -1 \\ 0 \end{pmatrix}, \; r \in \mathbb{R}$$

If you are given both planes in HNF, then you will have two equations for three unknowns $x_1$, $x_2$ and $x_3$. Then you solve the linear system and, in the generic case, you obtain $\mathbf{x}$ as a function of one parameter, i.e. the parametric form of the line of intersection.

By substitution you can check that the line lies in both planes.

$$\frac{1}{\sqrt{3}} \begin{pmatrix} 1 \\ 1 \\ 1 \end{pmatrix} \cdot \mathbf{x} - \frac{1}{\sqrt{3}} = 0$$

$$\frac{1}{\sqrt{2}} \begin{pmatrix} 1 \\ 1 \\ 0 \end{pmatrix} \cdot \mathbf{x} - \frac{1}{\sqrt{2}} = 0$$

$$\Rightarrow x_1 + x_2 + x_3 = 1, \; x_1 + x_2 = 1$$
$$\Rightarrow x_1 = r, \; x_2 = 1 - r, \; x_3 = 0$$

$$\Rightarrow \mathbf{x} = \begin{pmatrix} 0 \\ 1 \\ 0 \end{pmatrix} + r \begin{pmatrix} 1 \\ -1 \\ 0 \end{pmatrix}, \; r \in \mathbb{R}$$

# How do you calculate the distance from a point to a line or a plane?

First: By distance, we always mean the shortest distance, i.e. the one that is perpendicular to the line or the plane.

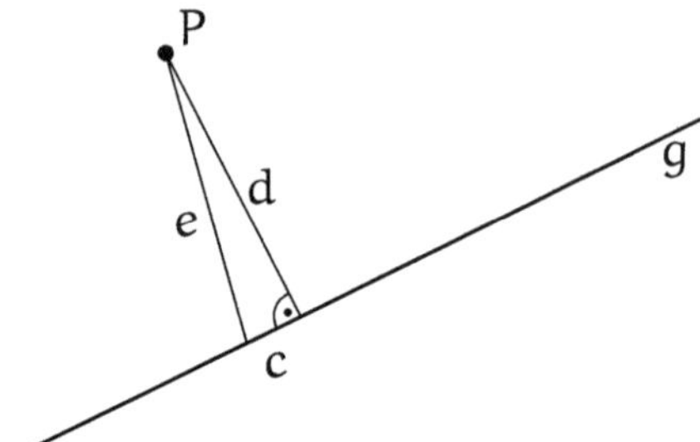


For $c > 0$, we have $d < \sqrt{d^2 + c^2} = e$.

Using the Hesse normal form, you can easily calculate the distance from a point to a line in 2D or from a point to a plane in 3D.

For the HNF by construction, $\mathbf{n} \cdot \mathbf{x} - d = 0$ holds for the points on the line or plane. If you plug in any point $\mathbf{x}$, then $|\mathbf{n} \cdot \mathbf{x} - d|$ gives you the distance. Check the derivation of the HNF again; the explanation is exactly the same.

Important for distance calculations is that in the HNF the normal vector is normalized, i.e. has length 1.

Distance from the point $(1, 1, 1)$ to the plane with intercepts $1, 1/2, 1/2$:

Intercept form: $x_1 + 2x_2 + 2x_3 = 1$

$$\text{HNF:} \quad \frac{1}{3}\begin{pmatrix} 1 \\ 2 \\ 2 \end{pmatrix} \cdot \mathbf{x} - \frac{1}{3} = 0$$

$$d(P, \mathbf{x}) = |\mathbf{n} \cdot \mathbf{x} - d|$$

$$= \left| \frac{1}{3}\begin{pmatrix} 1 \\ 2 \\ 2 \end{pmatrix} \cdot \mathbf{x} - \frac{1}{3} \right|$$

$$= \left| \frac{1}{3}\begin{pmatrix} 1 \\ 2 \\ 2 \end{pmatrix} \cdot \begin{pmatrix} 0 \\ 1 \\ 1 \end{pmatrix} - \frac{1}{3} \right| = 1$$

I will explain the calculation of the distance from a point to a line in 3D on the next page.

# How do you calculate the distances from a point to a line in 3D

The distance from a point to a line in 3D is calculated using the cross product.

You can derive the formula if you remember that the length of the cross product of two vectors is equal to the product of the lengths of both vectors multiplied by the absolute value of the sine of the included angle.

Distance from $\mathbf{x}_P$ to $g$: $\mathbf{x} = \mathbf{x}_0 + r\mathbf{a}$:

$$d = \frac{|(\mathbf{x}_P - \mathbf{x}_0) \times \mathbf{a}|}{|\mathbf{a}|}$$

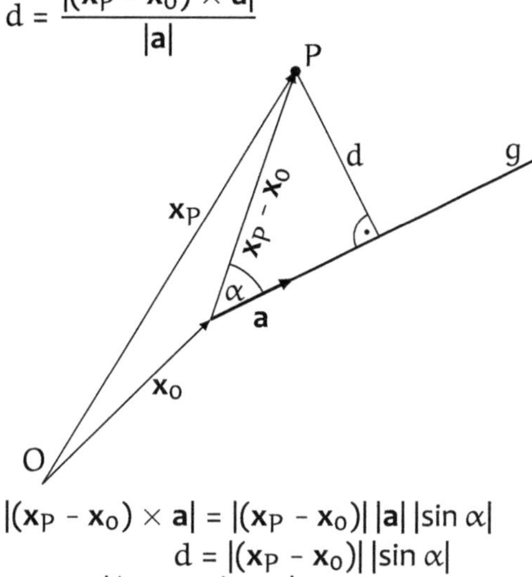


$$|(\mathbf{x}_P - \mathbf{x}_0) \times \mathbf{a}| = |(\mathbf{x}_P - \mathbf{x}_0)|\,|\mathbf{a}|\,|\sin \alpha|$$
$$d = |(\mathbf{x}_P - \mathbf{x}_0)|\,|\sin \alpha|$$
$$\Rightarrow d = \frac{|(\mathbf{x}_P - \mathbf{x}_0) \times \mathbf{a}|}{|\mathbf{a}|}$$

If the distance from the point to the line is zero, the point lies on the line. In this way, you obtain an equation for the line in 3D, similar to the normal form in 2D.

Line through $\mathbf{x}_0$ with direction vector $\mathbf{a}$ gives the Plücker form:
$$(\mathbf{x} - \mathbf{x}_0) \times \mathbf{a} = 0$$
$$\Rightarrow \mathbf{x} \times \mathbf{a} = \mathbf{x}_0 \times \mathbf{a}$$
You can also obtain this equation simply by taking the cross product of the parametric form with the direction vector.

You can also calculate the distance in another way, i.e. by finding the minimum of the distance $d(r)$ from the given point $\mathbf{x}_P$ to an arbitrary point on the line $\mathbf{x}(r) = \mathbf{x}_0 + r\mathbf{a}$. When $d(r)$ is minimal, $d^2(r)$ is also minimal, which is easier to calculate.

The square of the distance $d^2(r) = |\mathbf{x}_P - \mathbf{x}(r)|^2$ can be calculated using the Pythagorean theorem, resulting in a quadratic function in $r$, whose minimum you find by completing the square or by differentiation. The distance is then the square root of $d^2(r_{min})$.

# How do you calculate the distance between two lines, two planes, or a line and a plane?

You can calculate the distance between two skew lines by considering a direction perpendicular to both lines.

A direction perpendicular to both lines can be found by taking the cross product of the two direction vectors of the lines.

You obtain the distance by projecting the difference vector from any point on the first line to any point on the second line onto the direction perpendicular to both lines. The simplest way is to use the given points of the parametric equations, $x_A$ and $x_B$, for the two points.

If the two direction vectors are parallel, the two lines are parallel. In this case, the easiest way is to take any point on the first line and determine its distance to the second line as described on the previous page. The distance between the lines is then this distance.

You determine the distance from a line to a parallel plane by calculating the distance from any point on the line to the plane.

If you draw the perpendicular to both lines, and you look at it from the side, the two lines appear parallel, so the projection gives their distance.

If $x_A + r\,\mathbf{a}$ and $x_B + r\,\mathbf{b}$ are the two lines, then $\mathbf{n} = \mathbf{a} \times \mathbf{b}$ is a direction perpendicular to both lines.

$$d = (x_A - x_B) \cdot \frac{\mathbf{n}}{|\mathbf{n}|}$$

$$d = \frac{(x_A - x_B) \cdot (\mathbf{a} \times \mathbf{b})}{|\mathbf{a} \times \mathbf{b}|}$$

You can determine the distance between two parallel planes by taking any point on one of the planes and calculating its distance to the other plane. The distance between the planes is then this distance.

# Overview: Relative positions of points, lines, and planes

| Point | and point: | - are identical (distance = 0)<br>- **are different** (distance > 0) |
|---|---|---|
| | and line: | - point on the line (distance = 0)<br>- **point not on the line** (distance > 0) |
| | and plane: | - point on the plane (distance = 0)<br>- **point not on the plane**[3d] (distance > 0) |
| Line | and line: | - are identical (directions $\parallel$, distance = 0)<br>- are parallel (directions $\parallel$, distance > 0)<br>- **intersect** (directions $\nparallel$, distance = 0)<br>- **are skew**[3d] (directions $\nparallel$, distance > 0) |
| | and plane: | - line lies on the plane (= plane contains line)<br>- are parallel[3d] (line dir. $\in$ plane dir., distance > 0)<br>- **intersect**[3d] (line dir. $\notin$ plane dir., distance = 0) |
| Plane | and plane: | - identical (directions $\parallel$, distance = 0)<br>- parallel[3d] (directions $\parallel$, distance > 0)<br>- **intersect**[3d] (directions $\nparallel$) |

**Notes:**

Superscript [3d] means that this case only occurs in three dimensions.

Parallel direction means that the direction vectors span the same vector space. Line direction $\in$ plane direction means that the vector space of the line direction is a subspace of the plane direction.

The "generic position" is highlighted in bold above. The generic position of objects is the one that is stable under small changes, or, in other words, occurs with probability 1 when chosen at random.

For line and line, in two dimensions, the generic position is that they intersect; in three dimensions, that they are skew to each other.

# Overview: Products of vectors in different notations

| | Vector notation | Matrix notation with components | Meaning and application |
|---|---|---|---|
| **Dot product** | $\mathbf{a} \cdot \mathbf{b} \in \mathbb{R}$ where $\mathbf{a}, \mathbf{b} \in \mathbb{R}^n$ | $\mathbf{a}^\top \mathbf{b}$ <br> $(1 \times n)(n \times 1) = (1 \times 1)$ <br> $\mathbf{a}^\top \mathbf{b} = \sum_i a_i b_i$ | $\|a\|\|b\| \cos(\angle(\mathbf{a}, \mathbf{b})) =$ <br> $= 0$ if $\mathbf{a} \perp \mathbf{b}$. <br> angle calculation |
| **Cross or vector product** | $\mathbf{a} \times \mathbf{b} \in \mathbb{R}^3$ where $\mathbf{a}, \mathbf{b} \in \mathbb{R}^3$ | $M_\mathbf{a} \mathbf{b}$ <br> where <br> $M_\mathbf{a} = \begin{pmatrix} 0 & -a_3 & a_2 \\ a_3 & 0 & -a_1 \\ -a_2 & a_1 & 0 \end{pmatrix}$ <br> $(\mathbf{a} \times \mathbf{b})_i =$ <br> $= \sum_{j,k} \epsilon_{ijk} a_j b_k$ | $\mathbf{a} \times \mathbf{b} \perp \mathbf{a}, \mathbf{b}$ <br> $\|\mathbf{a} \times \mathbf{b}\| =$ <br> $= \|a\|\|b\| \|\sin(\angle(\mathbf{a}, \mathbf{b}))\|$ <br> $= 0$ if $\mathbf{a} \parallel \mathbf{b}$. <br> area calculation |
| **Dyadic product** | $\mathbf{a} \otimes \mathbf{b} \in \mathbb{R}^{n \times n}$ where $\mathbf{a}, \mathbf{b} \in \mathbb{R}^n$ | $\mathbf{a}\mathbf{b}^\top$ <br> $(n \times 1)(1 \times n) = (n \times n)$ <br> $(\mathbf{a}\mathbf{b})_{ij} = a_i b_j$ | $(\mathbf{a}\mathbf{b}^\top)\mathbf{c} = \mathbf{a}(\mathbf{b}^\top \mathbf{c}) \parallel \mathbf{a}$ <br> $\mathbf{a}\mathbf{a}^\top$ with $\|a\| = 1$ <br> projects onto $\mathbf{a}$. |

In one of the notations for the cross product, the $\epsilon$-tensor is used, which is zero if two indices are equal, +1 for the indices 123 and their cyclic permutations, and otherwise –1:

$$\epsilon_{ijk} = \begin{cases} +1 & (ijk) = (123), (231), (312) \\ -1 & (ijk) = (132), (213), (321) \\ 0 & \text{if two indices are equal.} \end{cases}$$

# A Glimpse into Probability Theory

Probability theory has its origins in the analysis of games of chance in the 17$^\text{th}$ century.

It was placed on a solid mathematical foundation when Kolmogorov, in the early 1930s, did not concern himself with defining what probability is, but instead formulated axioms that probability satisfies and from which calculation rules and mathematical theorems can be derived. Random variables then become certain maps.

I will explain this to you, as well as what the expected value and variance of random variables mean and how to calculate them.

You will get to know the binomial distribution and the Gaussian normal distribution as important distribution functions.

I will outline and explain a proof that shows you why many random variables, in which many random effects are superimposed additively, are approximately normally distributed.



A. Gründers, *Math Made Clear: From the Basics to Calculus*, https://doi.org/10.1007/978-3-662-73221-2_13

# What are the events of a random experiment and how can you calculate with them?

In a random experiment, there is a set of possible outcomes, which is called the sample space and often denoted by a capital omega, $\Omega$.

For a dice roll:
$$\Omega = \{1, 2, 3, 4, 5, 6\}$$

Any subset of this sample space is called an event. Any event that consists of only one outcome is called an elementary event.

Event $A$: even number
$A = \{2, 4, 6\}$
Event $B$: 5 or 6
$B = \{5, 6\}$

Since events are sets, you can calculate with them as with sets.

The union of two events occurs exactly when either one event or the other or both occur; this is called A or B and is written as $A \cup B$.

$$A \cup B = \{2, 4, 5, 6\}$$

The intersection of two events occurs exactly when both events occur; this is called A and B and is written as $A \cap B$.

$$A \cap B = \{6\}$$

The complementary event occurs exactly when the original event does not occur. This is called the complement of $A$ and is written as $\overline{A}$.

$$\overline{A} = \{1, 3, 5\}$$

The set $\overline{A}$ is the complement of the set A, i.e. the set of all elements from the sample space $\Omega$ that are not in A: $\overline{A} = \Omega \setminus A$.

# What is probability and how do you calculate unknown probabilities from known ones?

A probability is, at first, nothing more than a map that assigns to each event a number between zero and one.

$$P : \text{set of all events} \rightarrow [0, 1]$$
$$A \mapsto P(A)$$

The empty event is assigned zero, the certain event is assigned one.

$$P(\{\}) = 0$$
$$P(\Omega) = 1$$

The disjoint union of two events is assigned the sum of the probabilities.

$$P(A \cup B) = P(A) + P(B)$$
$$\text{for } A \cap B = \{\}$$

If all elementary events are equally likely, the probability of an elementary event is 1/number of events.

For a fair die:
$$P(\{1\}) = P(\{2\}) = P(\{3\}) = \ldots = \tfrac{1}{6}$$
$$P(\{2, 4, 6\}) = \tfrac{1}{6} + \tfrac{1}{6} + \tfrac{1}{6} = \tfrac{3}{6} = \tfrac{1}{2}$$

The complementary event $\overline{A}$ is assigned the probability one minus the probability of $A$.

$$P(\overline{A}) = 1 - P(A)$$
$$P(\overline{\{1\}}) = P(\{2, 3, 4, 5, 6\}) = \tfrac{5}{6}$$

This allows you to calculate unknown probabilities from known ones.

The probability of the union of two arbitrary events is obtained by adding their respective probabilities and subtracting the probability that both events occur.

$$P(A \cup B) = P(A) + \underbrace{P(B) - P(A \cap B)}_{P(B \backslash A)}$$

$A \cup B$ is the disjoint union of A and B \ A.

If two events are independent of each other, the probability of both events occurring is the product of the two probabilities.

$$P(A \cap B) = P(A)\,P(B)$$
for independent events

This can be seen as the definition of independent events.

# What does conditional probability mean?

The conditional probability describes the probability of an event under the condition that another event also occurs.

This means we only consider the events in which both A and B occur.

In order for the sum of the conditional probabilities for the occurrence of A to be one again, the combined probability for A and B must be divided by the probability of B.

For independent events, the conditional probability is equal to the original probability, and the product rule results again.

If we use the symmetry of intersection, i.e. $A \cap B = B \cap A$, and consider the other event as the condition, we obtain Bayes' theorem.

Often, is helpful to write the probability in the denominator as the sum of the probabilities over disjoint events.

For example, if D means that a person has a disease, and T means that a test indicates the person has the disease, then $P_T(D)$ can be calculated from $P_D(T)$ and further information.

$P_B(A)$ conditional probability: probability of event B under the condition A.

Also written as $P(A|B)$.

$$P_B(A) = \frac{P(A \cap B)}{P(B)}$$

$$P_B(A) = \frac{P(A \cap B)}{P(B)} = P(A)$$

$$\Rightarrow P(A \cap B) = P(A)P(B)$$
for independent events

$$P_B(A) = \frac{P(B \cap A)}{P(A)} = \frac{P_A(B)P(B)}{P(A)}$$

$$P_B(A) = \frac{P_A(B)P(A)}{P_A(B)P(A) + P_{\overline{A}}(B)P(\overline{A})}$$

Additionally you need the prevalence of the disease occurs and the sensitivity of the test. For further information search for "sensitivity" and "specificity".

# How can you calculate the probability from the number of possible events?

If the sample space $\Omega$ is finite and all elementary events are equally likely, each outcome has the probability 1/number of outcomes.

$$|\Omega| = n < \infty$$
$$\Rightarrow P(\text{elementary event}) = \frac{1}{|\Omega|} = \frac{1}{n}$$

Fair die, $|\Omega| = 6$:
$$\Rightarrow P(\{1\}) = P(\{2\}) = P(\{3\}) = \ldots = \frac{1}{6}$$

In such an experiment, the probability of an event is the ratio of the number of favorable outcomes of the event to the total number of all possible outcomes.

$$P(A) = \frac{\text{number of favorable cases}}{\text{number of all cases}}$$

$$P(A) = \frac{|A|}{|\Omega|}$$

In simple cases, the number of outcomes can be calculated by multiplication or using binomial coefficients, e.g. for 2 sixes in 5 rolls.

Number of all cases for 5 rolls:
$$|\Omega| = 6 \cdot 6 \cdot 6 \cdot 6 \cdot 6 = 6^5$$

Number of cases with 2 sixes:
For 5 rolls, there are $\binom{5}{2} = 10$ possibilities for the arrangement of the 2 sixes (xxx66, xx6x6, xx66x, x6xx6, x6x6x, ..., 66xxx).
For the 3 non-sixes, there are 5 possibilities each (1,2,3,4,5), so $5 \cdot 5 \cdot 5 = 5^3$ possibilities.

Thus, there are a total of $|A| = 10 \cdot 5^3$ favorable cases.

Thus, the probability is calculated as the ratio of number of the favorable cases to the number of all cases.

$$P = \frac{|A|}{|\Omega|} = \frac{10 \cdot 5^3}{6^5} \approx 16.1\%$$

# What is a random variable and its distribution?

A (real) random variable is a map from the sample space $\Omega$ to the real numbers.

$$X : \Omega \to \mathbb{R}$$
$$\omega \mapsto X(\omega)$$

Random variables describe the properties of an event that are of interest.

$\Omega$ = set of students = {Lisa, ...}
X: ID number, $X(\text{Lisa}) = 123456$
Y: exam grade, $Y(\text{Lisa}) = 95\%$

If $\Omega$ is a discrete set, the random variable will also only take values from a finite or countably infinite set.

These values are often denoted by $x_i$. The probability for $X = x_i$ is given by the sum of the probabilities of the elementary events $\omega$ with $X(\omega) = x_i$.

$x_i = x \in \mathbb{R}$, if $\omega \in \Omega$ exists
with $X(\omega) = x_i$.

$$p_i = P(X = x_i) = \sum_{X(\omega)=x_i} P(\omega)$$

The map is called the probability distribution of the random variable X.

$$P : \{x_i | i\} \to [0, 1]$$
$$x_i \mapsto p_i$$

For example, when tossing two coins, the number of coins X that show heads is a random variable.

There are four equally likely elementary events:
$$P(HH) = P(HT) = P(TH) = P(TT) = \tfrac{1}{4}$$

The distribution of X is obtained by considering which elementary events lead to which values of X.

$$P(\{X = 0\}) = P(\{TT\}) = \tfrac{1}{4}$$
$$P(\{X = 1\}) = P(\{HT, TH\}) = \tfrac{1}{4} + \tfrac{1}{4} = \tfrac{1}{2}$$
$$P(\{X = 2\}) = P(\{HH\}) = \tfrac{1}{4}$$

You can perform calculations with random variables. For example, you can calculate their expected value and their variance, you can form the sum or product of random variables, and much more.

# What are examples of probability distributions of discrete random variables?

A discrete random variable is a random variable that can take only finitely many or countably infinitely many values.

A random variable that can take arbitrary real values is not a discrete random variable.

Number of dots when rolling a die: finitely many values: $\{1, 2, 3, 4, 5, 6\}$

Number of rolls until a six appears: Countably infinitely many values: $\{1, 2, 3, \ldots\} = \mathbb{N}$

The probability distribution indicates how the total probability 1 is distributed among the values of the random variable.

$$p_i = P(X = x_i),$$
where $i$ takes finitely many or countably infinitely many values

The sum of them must always be 1.

$$\sum_i p_i = \sum_i P(X = x_i) = 1$$

The number of heads in $n$ tosses of a fair coin is a random variable with the probability distribution given on the right.

$$P_n(X = k) = \frac{1}{2^n}\binom{n}{k}, \, k = 0, 1, \ldots, n$$

$$\sum_{k=0}^{n} \frac{1}{2^n}\binom{n}{k} = \sum_{k=0}^{n} \frac{1}{2^n}\binom{n}{k} 1^k 1^{n-k}$$

$$= \frac{1}{2^n}(1 + 1)^n = 1$$

An example of a discrete random variable with infinitely many values is the number of random experiments needed for an event with probability $p$ to occur for the first time.

$P(X = n) = p(1 - p)^{n-1}, n = 1, 2, \ldots$
is the probability that an event with probability $p$ occurs for the first time on the $n$-th trial.

For $p = 1/6$, for example, this gives the number of rolls until the first six appears.

The sum of the probabilities over the different (here infinitely many) values of the random variable is also 1 in this case.

$$\sum_{n=1}^{\infty} p(1 - p)^{n-1} = p \sum_{n=1}^{\infty} (1 - p)^{n-1}$$

$$= p \cdot \frac{1}{1 - (1 - p)} = 1$$

# How do you add random variables?

To calculate the probability for the sum S of the dots of two dice, you consider in which ways a given sum can occur and add the respective probabilities, since the corresponding events are disjoint.

In general, the probability for $S = k$ is given as the sum of the probabilities of all events $(D_1 = k_1, D_2 = k_2)$ with $k_1 + k_2 = k$.

In this way, you can calculate the probability for other sums of number of dots. Doing so, you obtain as the sum of two discrete uniform distributions, the discrete triangular distribution shown on the right.

$$\{S = 4\} = \{(D_1 = 1, D_2 = 3)\}$$
$$\cup \{(D_1 = 2, D_2 = 2)\}$$
$$\cup \{(D_1 = 3, D_2 = 1)\}$$
$$\Rightarrow P(S = 4) = \tfrac{1}{36} + \tfrac{1}{36} + \tfrac{1}{36} = \tfrac{1}{12}$$

$$P(S = k) = \sum_{k_1 + k_2 = k} P(D_1 = k_1, D_2 = k_2)$$
$$= \sum_{k_1} P(D_1 = k_1, D_2 = k - k_1)$$

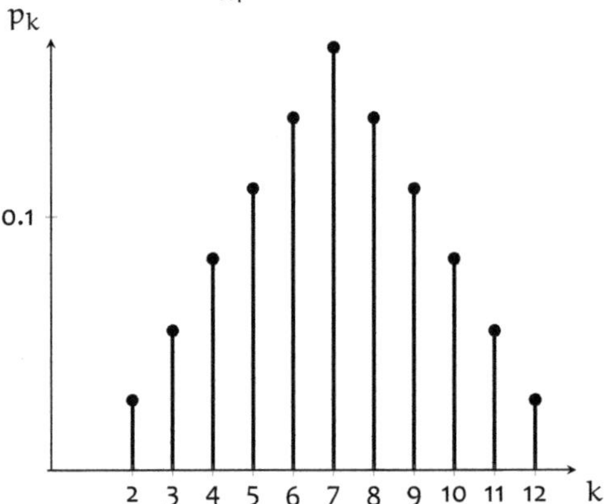

Probability $p_k$ for the sum of number of dots $S = k$ with two dice

Similarly the number of successes $S_n$ in $n$ stochastically independent repetitions of a Bernoulli trial is the sum of the individual random variables.

This yields the binomial distribution, which will be explained to you later.

Bernoulli random variable X:
$$P(X = 1) = p, \quad P(X = 0) = 1 - p$$

Binomial distribution $S_n$
$$S_n = X_1 + X_2 + \cdots + X_n,$$
where $X_i$ are independent and identically distributed Bernoulli variables.

$S_n$ has a binomial distribution with parameters $n$ and $p$.

# What is the expected value of a random variable?

The expected value of a random variable is the sum of the values of the random variable multiplied by their probabilities.

If a random variable takes two values, each with probability 50%, the expected value is equal to the arithmetic mean of the two values.

If you imagine the probability distribution as a histogram with bars carrying mass, the expected value is the $x$-value of the center of mass.

The expected value is not the value you should expect as the outcome of the random experiment; it does not even have to be a possible value of the random variable.

It can be shown that the mean value $\overline{X}^{(n)}$ in $n$ repetitions of a random experiment converges stochastically to the expected value as $n \to \infty$.

$$E(X) = \sum_i P(X = x_i)\, x_i = \sum_i p_i x_i$$

$E(X)$ is also called $\mu_X$.

$$E(X) = \sum_i p_i x_i$$
$$= 0.5 x_1 + 0.5 x_2$$
$$= \frac{1}{2}(x_1 + x_2)$$

The expected value is therefore the point where you have to place your finger under the histogram for it to be balanced.

The expected value when rolling a fair die is:

$$E(X) = \frac{1}{6} \cdot 1 + \frac{1}{6} \cdot 2 + \frac{1}{6} \cdot 3 + \frac{1}{6} \cdot 4+$$
$$+ \frac{1}{6} \cdot 5 + \frac{1}{6} \cdot 6 = \frac{1}{6} \cdot \frac{7 \cdot 6}{2} = 3.5.$$

Let $X_i$ be stochastically independent and identically distributed random variables with finite variance, then for the mean $\overline{X}^{(n)} = \frac{1}{n} \sum_{i=1}^{n} X_i$

it holds
$$\lim_{n \to \infty} P(|\overline{X}^{(n)} - \mu_X| > \epsilon) = 0$$

for every $\epsilon > 0$.

This is called the weak law of large numbers.

There is also a strong law of large numbers, which states that $\overline{X}^{(n)}$ converges almost surely to $\mu_X$ as $n \to \infty$. We will not touch on it.

# What is the variance of a random variable?

The variance is a measure of the spread of a random variable.

It is the expected value of the square of the deviation from the expected value.

$$V(X) = E((X - E(X))^2)$$

With the abbreviation $\mu = E(X)$:

$$V(X) = E((X - \mu)^2)$$
$$= \sum_i (x_i - \mu)^2 P(X = x_i)$$

$$V(X) = E((X - \mu)^2)$$
$$= E(X^2 - 2\mu X + \mu^2)$$
$$= E(X^2) - 2\mu \underbrace{E(X)}_{\mu} + \mu^2$$
$$= E(X^2) - \mu^2$$

If you expand the square and note that the expected value is linear and $E(X) = \mu$, you see that the variance is the difference between the expected value of $X^2$ and the square of the expected value, $\mu^2$.

If you multiply a random variable by a factor, the variance is multiplied by the square of the factor.

$$V(aX) = a^2 V(X)$$

The square root of the variance of a random variable is called its standard deviation $\sigma$. Like the variance, it is always positive and scales linearly with the random variable.

$$\sigma_X = \sqrt{V(X)}$$

$$\sigma_{aX} = |a|\, \sigma_X$$

For independent random variables, the expected value of the product is equal to the product of the expected values.

For independent random variables:
$E(XY) = E(X)E(Y)$.
This results from the definition of the expected value and the defining equation of stochastic independence:
$$P(X = x_i, Y = y_j) = P(X = x_i)P(Y = y_j).$$

From this results the additivity of the variance in the case of independence.

We get $V(X + Y) = V(X) + V(Y)$
for independent random variables.
We will not prove this here.

# How are the expected value and variance of the mean value of $n$ random variables calculated?

For the sum of identically distributed random variables, the means add up.

$$E(X_1 + X_2 + \ldots + X_n)$$
$$= E(X_1) + E(X_2) + \ldots + E(X_n) = nE(X)$$

The expected value is linear.

$$E(aX) = aE(x)$$

Thus, the expected value of the mean value is equal to the expected value of the individual random variables.

$$E(\overline{X}) = E\left(\frac{1}{n}\sum_{i=1}^{n} X_i\right) = \frac{1}{n}nE(X) = E(X)$$
$$\mu_{\overline{X}} = \mu_X$$

For the sum of independent and identically distributed random variables, the variance values add up.

$$V(X_1 + X_2 + \ldots + X_n)$$
$$= V(X_1) + V(X_2) + \ldots + V(X_n) = nV(x)$$

The variance scales quadratically.

$$V(aX) = a^2 V(X)$$

Thus, the variance of the mean value can be calculated as shown on the right.

$$V(\overline{X}) = V\left(\frac{1}{n}\sum_{i=1}^{n} X_i\right) = \frac{1}{n^2}nV(X) = \frac{V(X)}{n}$$

The standard deviation of a random variable is the square root of the variance.

$$\sigma_{\overline{X}} = \sqrt{V(\overline{X})} = \frac{\sqrt{V(x)}}{\sqrt{n}} = \frac{\sigma_X}{\sqrt{n}}$$

Therefore, the mean value has the same expected value as the random variable itself, but fluctuates much less than it.

That is why measurements are repeated. However, to reduce the standard deviation by a factor of 10, you need to make 100 measurements.

With the decrease of the variance of the mean value for many repetitions, the weak law of large numbers can be proved.

# What do you need the binomial distribution for?

The binomial distribution $B_{n,p}(k)$ gives the probability of k successes when performing a random experiment n times and the probability of a success in a single trial is p. The individual trials are assumed to be independent of each other.

$B_{5,1/6}(k) =$
Probability of getting k sixes when rolling a fair die 5 times.

$n = 5$: number of repetitions
$p = 1/6$: probability of getting a
six in a single roll

You calculate the probability of k successes by first considering that the probability of a fixed order of k successes and thus $n - k$ failures is the product of the individual probabilities.

For 5 rolls, the probability of getting 2 sixes and 3 non-sixes in that order is:
$p(\text{"66xxx"}) = \left(\frac{1}{6}\right)^2 \left(1 - \frac{1}{6}\right)^3.$

In general, for a fixed order of k successes in n repetitions:
$$p(\underbrace{A\ldots A}_{k}\,\underbrace{\overline{A}\ldots\overline{A}}_{n-k}) = p^k(1 - p)^{n-k}$$

The total probability is obtained by considering that there are $\binom{n}{k}$ possible orders.

For $k = 2$ sixes and $n - k = 5 - 2$ non-sixes, there are $\binom{5}{2} = \frac{5 \cdot 4}{1 \cdot 2} = 10$ possible orders:

66xxx, 6x6xx, 6xx6x, 6xxx6, x66xx,

x6x6x, x6xx6, xx66x, xx6x6, xxx66.

The probability is the same in each case.

$p(\text{"66xxx"}) = p(\text{"6x6xx"}) = \ldots =$
$= \left(\frac{1}{6}\right)^2 \left(\frac{5}{6}\right)^3$

By multiplying the individual probability by the number of different possibilities, you obtain the total probability of k successes.

Probability of 2 sixes in 5 rolls:
$p = \binom{5}{2} \left(\frac{1}{6}\right)^2 \left(\frac{5}{6}\right)^3 = 10 \cdot \frac{5^3}{6^5} = \frac{625}{3888}$
$\approx 0.161$

This probability distribution is called the binomial distribution. It is defined for fixed $p \in (0, 1)$, $n \in \mathbb{N}$ and $k = 0, \ldots, n$.

The probability of k successes in n repetitions is:

$B_{n,p}(k) = \binom{n}{k}p^k(1 - p)^{n-k}.$

# What does the binomial distribution look like?

The binomial distribution gives the probability for $k$ successes in $n$ trials with individual probability $p$. The value of $k$ takes on $n + 1$ integer values, i.e. $k = 0, 1, \ldots, n$. Here you can see four diagrams for $n = 6$ and different values of $p$. On the right, $p = 1/6$.

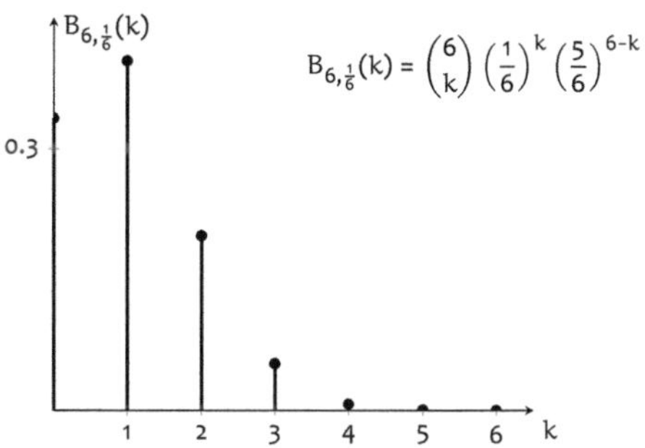



Now on the right, $p = 1/3$. The probability for $k$ successes in 6 trials, with individual probability $p = 1/3$, is maximal for $k = 2$. For $k = 1$ it is higher than for $k = 3$.

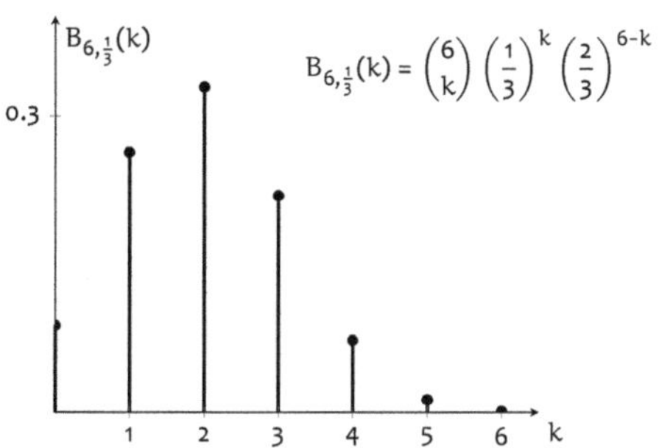



For $p = 1/2$, the probability distribution for $k$ is symmetric about the expected value $n \cdot p = n/2$. In the diagram it is symmetric about $k = 3$.

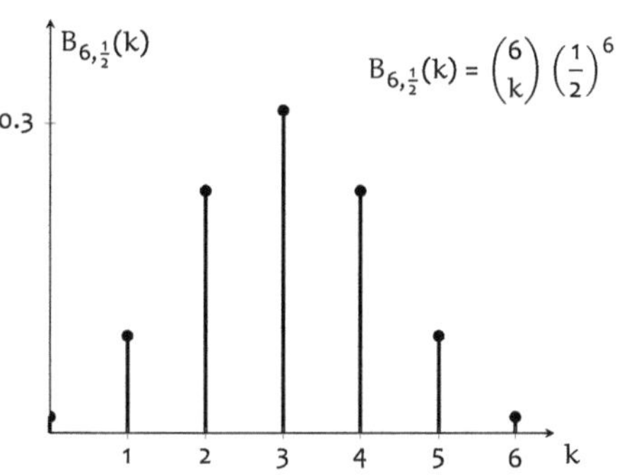



For $p' = 1 - p$, the probability distribution is the one for $p$, but mirrored about $n/2$. On the right, you can see it for $p' = 1 - \frac{1}{3} = \frac{2}{3}$.

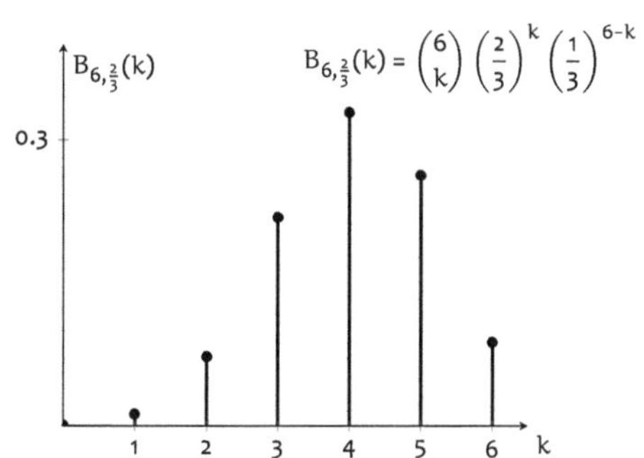

# What is a continuous probability distribution?

In contrast to before, we now consider continuous random variables, which can take not only discrete values, but, for example, all values from an interval, or even all real numbers.

The probability that the random variable, X, lies in the interval $[a, b]$ is equal to the area under the probability density function, f, between $x = a$ and $x = b$.

The probability density function (PDF), f, has the antiderivative F where $F(x) = \int_{-\infty}^{x} f(t)\,dt$. The function, $F(x)$, is called the cumulative distribution function.

The probability density, $f(x)$, is approximately equal to the probability that X takes a value in the interval $[x, x + \Delta x]$, divided by $\Delta x$, for small $\Delta x$.

In the limit $\Delta x \to 0$, you can see that the density $f(x)$ is the derivative of the cumulative distribution function.

For calculating normalization, mean, and variance – in the case of continuous random variables – you use integrals instead of sums.

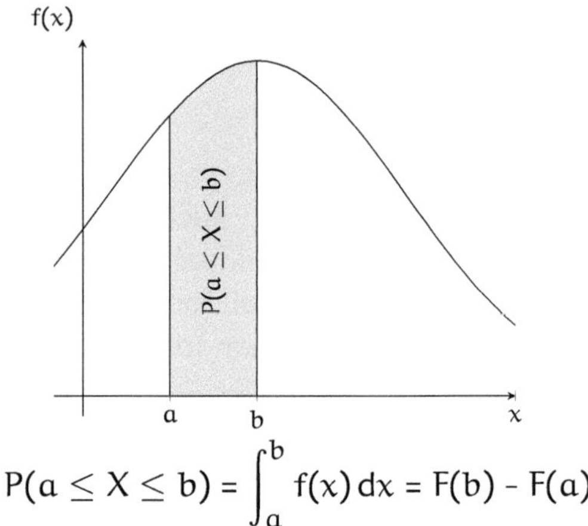



$$P(a \leq X \leq b) = \int_{a}^{b} f(x)\,dx = F(b) - F(a)$$

$$f(x) \approx \frac{P(x \leq X \leq x + \Delta x)}{\Delta x}$$

$$f(x) = \lim_{\Delta x \to 0} \frac{P(x \leq X \leq x + \Delta x)}{\Delta x}$$

$$= \lim_{\Delta x \to 0} \frac{F(x + \Delta x) - F(x)}{\Delta x} = F'(x)$$

$$1 = \int f(x)\,dx \qquad \left( \sum_i p_i \right)$$

$$E(X) = \int x f(x)\,dx \qquad \left( \sum_i x_i p_i \right)$$

$$V(X) = \int (x - \mu)^2 f(x)\,dx \quad \left( \sum_i (x_i - \mu)^2 p_i \right)$$

# What is the Gaussian normal distribution?

An important continuous probability distribution is the Gaussian normal distribution. In the simplest case, it has mean zero and standard deviation one.

$$\varphi(x) = \frac{1}{\sqrt{2\pi}} e^{-\frac{x^2}{2}}$$

density of the standard normal distribution with

$$\mu_X = 0,\ \sigma_X = 1$$

The prefactor $1/\sqrt{2\pi}$ is only needed for the density of the normal distribution to be normalized.

$$\int_{-\infty}^{\infty} e^{-\frac{x^2}{2}}\, dx = \sqrt{2\pi}$$

$$\Rightarrow \int_{-\infty}^{\infty} \frac{1}{\sqrt{2\pi}} e^{-\frac{x^2}{2}}\, dx = 1$$

If you shift the standard normal distribution to the right by $\mu$ and scale it by $\sigma$, the mean value becomes $\mu$ and the standard deviation $\sigma$.

$$\varphi_{\mu,\sigma}(x) = \frac{1}{\sigma\sqrt{2\pi}} e^{-\frac{1}{2}\left(\frac{x-\mu}{\sigma}\right)^2}$$

$$\mu_X = \mu,\ \sigma_X = \sigma$$

The cumulative distribution function $\Phi$ corresponding to $\varphi$, i.e. the antiderivative of $\varphi$, can't be expressed in terms of elementary functions.

$$\Phi(x) = \frac{1}{\sqrt{2\pi}} \int_{-\infty}^{x} e^{-\frac{x^2}{2}}\, dx$$

can only be calculated approximately.

The values $\Phi(-\infty)$, $\Phi(0)$, and $\Phi(\infty)$ can be obtained from the definition and from the normalization and symmetry of $\Phi$.

$$\Phi(-\infty) = 0,\ \Phi(\infty) = 1$$
$$\Phi(\infty) - \Phi(-\infty) = 1 \quad \text{(normalization)}$$
$$\Phi(0) = \frac{1}{2} \quad\quad\quad \text{(symmetry)}$$

Other values of $\Phi(x)$ can be approximated using series. Due to symmetry it's sufficient to consider $x \geq 0$.

From the symmetry $\varphi(-x) = \varphi(x)$ we get $\Phi(-x) = 1 - \Phi(x)$.

The probability that the random variable $x$ deviates from its mean value by at most $a$ is $P(|x| \leq a) =$
$= P(-a \leq x \leq a) = \Phi(a) - \Phi(-a) =$
$= \Phi(a) - (1 - \Phi(a)) = 2\Phi(a) - 1.$

$P(-3 \leq x \leq 3) = 2\Phi(3) - 1$
$\approx 2 \cdot 0.9865 - 1 \approx 99.7\%$ means that a normally distributed random variable takes a value between $\mu - 3\sigma$ and $\mu + 3\sigma$ with probability 99.7%.

# How can you understand the Gaussian normal distribution? (I)

The simplest random experiment has two outcomes, success ($x=1$) or failure ($x=0$), each with probability $\frac{1}{2}$, i.e. a Bernoulli trial with $p=\frac{1}{2}$.

If you perform $n$ independent Bernoulli trials $X_i, i = 1, \ldots, n$, the number of successes is equal to the sum of the individual random variables.

Thus, the simplest case for the sum of random variables is the binomial distribution with $p = 1/2$.

$$P(X = 0) = \tfrac{1}{2}, P(X = 1) = \tfrac{1}{2}$$

$$E(X) = \tfrac{1}{2} \cdot 0 + \tfrac{1}{2} \cdot 1 = \tfrac{1}{2}$$

$$V(X) = \tfrac{1}{2}(0 - \tfrac{1}{2})^2 + \tfrac{1}{2}(1 - \tfrac{1}{2})^2 = \tfrac{1}{4}$$

$$S_n = X_1 + X_2 + \ldots + X_n$$

$$p_n(k) = P(S_n = k) = \tfrac{1}{2^n} \binom{n}{k}$$

$$E(S_n) = nE(X) = \tfrac{n}{2}$$

$$V(S_n) = nV(X) = \tfrac{n}{4} \Rightarrow \sigma = \tfrac{\sqrt{n}}{2}$$

Except for the normalization, the probabilities are simply the binomial coefficients that you know from Pascal's triangle.

The binomial distribution is symmetric for $p = 1/2$ about $k = n/2$, where it reaches its maximum for even $n$, as shown on the right for $n = 6$.

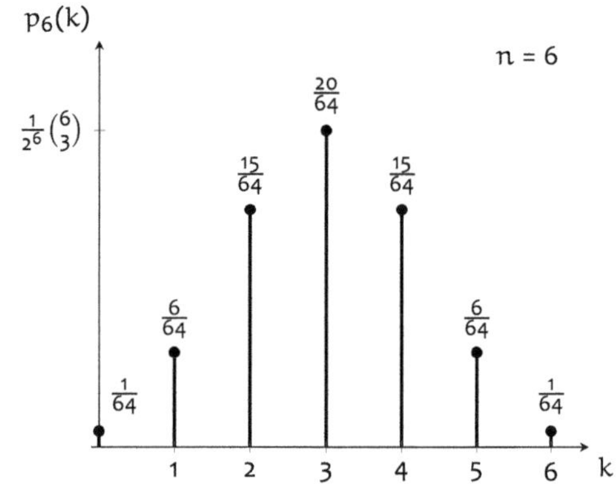


For large $n$, the distribution function approaches a normal distribution, as we will see on the next page. It can be shown that other sums of many independent random variables are also approximately normally distributed. This statement is known as the central limit theorem.

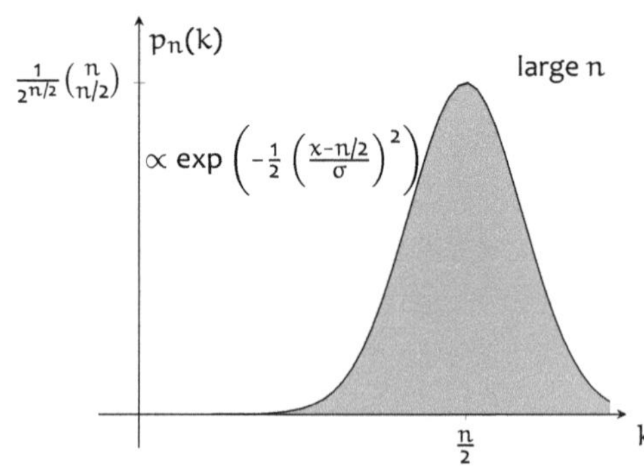


270

# How can you understand the Gaussian normal distribution (II)

The binomial distribution for $p = 1/2$ is symmetric about $k = n/2$, where it reaches its maximum for even $n$.

$$p_n(k) = \frac{1}{2^n}\binom{n}{k} = \frac{1}{2^n}\binom{n}{n-k} = p_n(n-k)$$

$$p_n(k = \tfrac{n}{2}) = \frac{1}{2^n}\binom{n}{n/2}$$

We introduce the variable $j = k - m$, which measures the distance from the mean value $m = n/2$.

$$p_n(m+j) = \frac{1}{2^{2m}}\binom{2m}{m+j} = p_n(m-j)$$

$$p_n(m) = \frac{1}{2^{2m}}\binom{2m}{m}$$

To determine the shape of the distribution for large $n$, we divide by the maximum value.

$$\frac{p_n(m+j)}{p_n(m)} = \frac{\binom{2m}{m+j}}{\binom{2m}{m}} = \frac{\frac{(2m)!}{(m+j)!\,(m-j)!}}{\frac{(2m)!}{(m!)^2}}$$

$$= \frac{(m!)^2}{(m+j)!\,(m-j)!}$$

For the factorials, we use Stirling's formula $n! \approx \sqrt{2\pi n}(n/e)^n$ for large $n$, which is already quite accurate for $n = 10$ with $3.59...\cdot 10^6$ vs. exactly $3.62...\cdot 10^6$.

You can see the main term as follows:
$\ln(n!) = \sum_{i=1}^{n}\ln(i) \approx \int_1^n \ln(x)dx = n(\ln(n)-1)$, so $n! \approx (n/e)^n$.

The square root factors can be omitted for our purposes; for large $m$ and small $j$, they are almost the same in the numerator and the denominator.

$$\frac{p_n(m+j)}{p_n(m)} \approx \frac{\left(\frac{m}{e}\right)^{2m}}{\left(\frac{m+j}{e}\right)^{m+j}\left(\frac{m-j}{e}\right)^{m-j}}$$

$$= \frac{m^{2m}}{(m+j)^{m+j}(m-j)^{m-j}} = \frac{1}{\left(1+\frac{j}{m}\right)^{m+j}\left(1-\frac{j}{m}\right)^{m-j}}$$

Thus you obtain a fraction without factorials, which you can further simplify.

Then we take the negative logarithm and use the approximation $\ln(1 + x) \approx x - x^2/2$ for small $x = j/m$, i.e. near the mean value.

$$-\ln\left(\frac{p_n(m+j)}{p_n(m)}\right) \approx (m+j)\ln\left(1+\frac{j}{m}\right)$$
$$+ (m-j)\ln\left(1-\frac{j}{m}\right)$$
$$\approx m\left((1+x)\ln(1+x) + (1-x)\ln(1-x)\right)$$
$$\approx m\left((1+x)(x-\tfrac{x^2}{2}) + (1-x)(-x-\tfrac{x^2}{2})\right)$$
$$\approx mx^2 + \text{higher order terms}$$

$$\approx m\left(\frac{j}{m}\right)^2 = \frac{j^2}{m}$$

Therefore, up to normalization, you obtain the normal distribution (see previous page).

$$p_n(m+j) \propto \exp\left(-\frac{j^2}{m}\right) = \exp\left(-\frac{2j^2}{n}\right)$$
$$= \exp\left(-\frac{1}{2}\left(\frac{j}{\sigma}\right)^2\right) \text{ with } \sigma = \frac{\sqrt{n}}{2}$$

# What's next? – Outlook

If you have enjoyed the book so far and want to know what comes next, I will give you a brief outlook in the following, starting with the question of whether there are different sizes of infinity.

Afterwards, I will show you, on the one hand, how to extend the real numbers so that you can also take square roots of negative numbers. This leads to the complex numbers, with which you can calculate in much the same way as with the real numbers.

On the other hand, I will show you how you can approximate a function even better than by the tangent by using higher powers and higher derivatives. This leads to the Taylor series, which can approximate many functions as closely as desired.

After that, I will combine both and explain to you a connection between $\pi$ and e that is often called the most beautiful formula in the world.

Finally, as a bonus, there will be a brief introduction to geometric algebra, a generalization of vector algebra that allows you to calculate with higher-dimensional geometric objects.

# Are all infinite sets the same size?

A set has finitely many elements or infinitely many.

$|\{-1, 0, 1\}| = 3$
$|\mathbb{N}| = |\{1, 2, 3, \ldots\}| > n$ for all $n \in \mathbb{N}$

Two sets are called equinumerous (i.e. having the same size) if there is a bijection between them.

$$\begin{array}{cccc} \{1, 2, & 3, & 4, & \ldots\} \\ \updownarrow \updownarrow & \updownarrow & \updownarrow & \\ \{1, 8, & 27, & 64, & \ldots\} \end{array}$$

The set of cube numbers is the same size as the set of natural numbers.

It can be shown, that there is a bijection between the set of rational numbers and the set of natural numbers.

The rational numbers (fractions) can be enumerated.
You can find out why this is the case by searching for "1$^{st}$ Cantor diagonal argument"

Thus, the set $\mathbb{Q}$ has the same cardinality as the set $\mathbb{N}$; the two sets are equally large.

$|\mathbb{N}| = |\mathbb{Q}|$

A set, which has the same cardinality as the natural numbers, is called countably infinite.

$\mathbb{N}$ is countably infinite
$\mathbb{Q}$ is countably infinite
You can enumerate the elements.

An infinite set that is not countable has a greater cardinality than $\mathbb{N}$; it is called uncountably infinite, or simply uncountable.

The elements of an uncountable set can't be enumerated; you can't create a complete list containing all elements.

It can be shown that the real numbers are uncountably infinite.

$|\mathbb{R}| > |\mathbb{N}|$
You can be find out why this is the case by searching for the "2$^{nd}$ Cantor diagonal argument"

With the axioms of the standard set theory (ZFC axioms), it is undecidable whether there is a cardinality between the set of natural numbers and the set of real numbers.

$|\mathbb{N}| < ? < |\mathbb{R}|$ "Continuum problem"
It can be proven that neither the assumption of the existence of an intermediate cardinality leads to a contradiction to the standard set theory, nor does the assumption of non-existence.

# What are quantifiers useful for?

With the symbols of propositional logic, you can formally write mathematical statements.

$$x^2 > 1 \Leftrightarrow (x < -1 \lor x > 1)$$

Predicate logic is an extension of propositional logic. It includes quantifiers which specify the scope of a logical form.

$\forall$ and $\exists$ are quantifiers

$\forall$ means "for all"
$\exists$ means "there exists"

The universal quantifier $\forall$ expresses the fact that a logical form holds for all values of a variable.

$\forall n \in \mathbb{Z} : n^2 \geq 0$
For all $n \in \mathbb{Z}$, $n^2 \geq 0$ holds.
You can imagine it with the logic connective $\wedge$:
$(0^2 \geq 0) \wedge (1^2 \geq 0) \wedge ((-1)^2 \geq 0) \ldots$ is true, because the statements are true for all $n$.

The existential quantifier $\exists$ expresses the fact that a statement holds for at least one variable; in other words, that there is at least one value of the variable for which the statement holds.

$\exists n \in \mathbb{Z} : n^2 = 1$
There is (at least) one $n \in \mathbb{Z}$ with $n^2 = 1$.
You can imagine it with the logic connective $\vee$:
$(0^2 = 1) \vee (1^2 = 1) \wedge ((-1)^2 = 1) \ldots$ is true, because there is at least one $n$ (here $n = 1$ or $n = -1$) for which the statement is true.

The order matters; in the natural numbers, "$\forall n \exists m : m = 2n$" is true: For every natural number ($n$), there is another natural number ($m$) that is twice as large as the first.

With quantifiers swapped:
In the natural numbers,
"$\exists m \forall n : m = 2n$" is false:

There is no natural number ($m$) that is twice as large as all other natural numbers ($n$), for all at the same time.

All definitions and theorems in this book can be written formally in this way.

The function f is continuous at $\hat{x}$. $\Leftrightarrow$
$\forall \epsilon > 0 \, \exists \delta > 0 \, \forall x$ with $|x - \hat{x}| < \delta$ :
$|f(x) - f(\hat{x})| < \epsilon$

You can formalize entire proofs with proof assistants. Then you receive immediate feedback on whether a proof is complete.

You can playfully get to know the proof assistant LEAN with the "Natural Number game" :
`https://adam.math.hhu.de`

# What is algebra?

By abstract algebra we understand the study of algebraic structures such as groups, rings, fields, which have been independent objects of investigation for a good century.

A group is a set with a binary associative operation that has a neutral element and for every element an inverse.

For a ring a second binary operation and further axioms are added; a field is a commutative ring with a multiplicative neutral element, in which every element except zero has a multiplicative inverse.

In particular, algebra deals with solving polynomial (= algebraic) equations and systems of equations and is the basis of number theory and algebraic geometry.

The question, which equations (for example, of degree five) can be solved with nested roots, is closely connected to the theory of groups that can be assigned to the equations.

Furthermore, an algebra is a specific mathematical structure, namely a vector space that is additionally equipped with a multiplication compatible with the vector space operations.

Elementary algebra refers to calculations, for example, in the field of rational or real numbers and solving simple equations for an unknown.

$(\mathbb{Z}, +)$ is a group.
$(2 + 3) + 5 = 2 + (3 + 5)$
$7 + 0 = 7$
$3 + (-3) = 0$

$(\mathbb{Z}, +, \cdot)$ is a ring.
$(\{m + n\sqrt{13} \mid m, n \in \mathbb{Z}\}, \cdot, +)$ is a ring.

$(\mathbb{R}, +, \cdot)$ is a field.
$(\{r + s\sqrt{13} \mid r, s \in \mathbb{Q}\}, \cdot, +)$ is a field.

$x^2 - 13y^2 = 1$
has as its smallest integer solution
$x = 649, y = 180$.

If you are interested in this, look up the keyword "Galois theory" on the internet.
This was the beginning of modern, abstract algebra.

The vector space $\mathbb{R}^3$ becomes an algebra with the exterior product.

# What do you need complex numbers for?

In order to reverse the operation of addition for all natural numbers, we constructed the integers.

$$a + x = b \quad \Rightarrow \quad x = b - a$$
$$\text{natural} \quad \rightarrow \quad \text{integer}$$
$$\text{numbers} \qquad\qquad \text{numbers}$$

For the reversal of multiplication, we introduced the rational numbers.

$$ax = b \quad \Rightarrow \quad x = \frac{b}{a}$$
$$\text{integer} \quad \rightarrow \quad \text{rational}$$
$$\text{numbers} \qquad\qquad \text{numbers}$$

Then we introduced the real numbers as sequences of rational numbers, such that we were able to extract a root from any positive number.

$$1.4 \qquad\qquad 1.4142135\ldots$$
$$1.41$$
$$1.414$$
$$1.4142$$
$$\vdots$$

$$x^2 = 2 \quad \Rightarrow \quad x = \sqrt{2} = 1.4142135\ldots$$
$$\text{rational} \quad \rightarrow \quad \text{real}$$
$$\text{numbers} \qquad\qquad \text{numbers}$$

However, we still can't extract a square root from negative numbers, since squaring any real number yields a number greater than or equal to zero.

$$x \in \mathbb{R} \Rightarrow x^2 \geq 0$$
$$\Rightarrow \text{There is no } x \text{ with } x^2 < 0.$$
$$\Rightarrow \sqrt{-1} \notin \mathbb{R}$$

To fix this, we just define i as a number whose square is -1. This i is called the imaginary unit.

$$i^2 = -1$$

Numbers that can be written as the sum of a real number and a real multiple of i are called complex numbers.

$$2 + 3i \in \mathbb{C}$$
$$\sqrt{2} - 4.567i \in \mathbb{C}$$

The set of all complex numbers is denoted by $\mathbb{C}$.

$$\mathbb{C} = \{a + bi \mid a, b \in \mathbb{R}\}$$

# How do you calculate with complex numbers?

You add or subtract two complex numbers by adding or subtracting their real parts and adding or subtracting their imaginary parts.

$$(a + bi) + (c + di) = (a + c) + (b + d)i$$
$$(a + bi) - (c + di) = (a - c) + (b - d)i$$
$$(3 + i) + (4 + 5i) = 7 + 6i$$

You multiply two complex numbers by expanding the brackets and then taking into account that $i^2 = -1$.

$$(a+bi)(c+di) = ac+ad\,i+bc\,i+bd\,i^2 = (ac - bd) + (ad + bc)i$$
$$(1 + i)(1 + 2i) = 1 + i + 2i + 2i^2 = -1 + 3i$$

For dividing by a complex number $c + di$, you use a trick: you multiply the numerator and denominator by $c - di$ so that the denominator becomes a real number.

$$\frac{a + bi}{c + di} = \frac{(a + bi)(c - di)}{(c + di)(c - di)}$$
$$= \frac{(ac - bd) + (bc - ad)i}{c^2 + d^2}$$
$$= \frac{ac - bd}{c^2 + d^2} + \frac{bc - ad}{c^2 + d^2}i$$

Here is a specific example.

$$\frac{1 + i}{1 - i} = \frac{(1 + i)(1 + i)}{(1 - i)(1 + i)} = \frac{1 + 2i - 1}{1 + 1} = i$$

The powers of a complex number can be formed using the binomial formulas. Doing so, you need the powers of i, which have a period of 4.

$$i^0 = 1, i^1 = i, i^2 = -1, i^3 = -i,$$
$$i^4 = 1, i^5 = i, \ldots$$

If you know Euler's formula, multiplication and division become much easier.

$$z_1 = r_1 e^{i\varphi_1}, z_2 = r_2 e^{i\varphi_2}$$
$$z_1 z_2 = r_1 r_2 e^{i(\varphi_1 + \varphi_2)}$$
$$\frac{z_1}{z_2} = \frac{r_1}{r_2} e^{i(\varphi_1 - \varphi_2)}$$

# How can you represent and add complex numbers on the plane?

If you draw a second number line perpendicular to the real number line, with the imaginary unit i and its real multiples, you get the complex plane, also called Gaussian plane. Each point corresponds to a complex number $a + bi$.

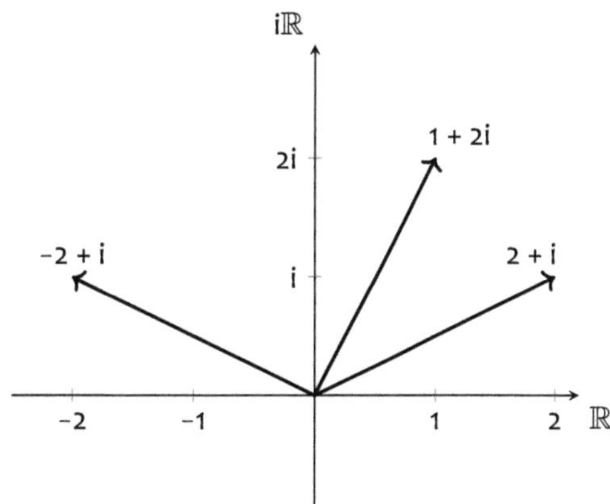


You can also think of the complex number $a + bi$ as an arrow from the origin to the corresponding point.

To do this, go $a$ units in the direction of the real axis $\mathbb{R}$ and $b$ units in the direction of the imaginary axis $i\mathbb{R}$.

You add two complex numbers by adding the respective coordinates; you can visualize this by placing the arrows head to tail.

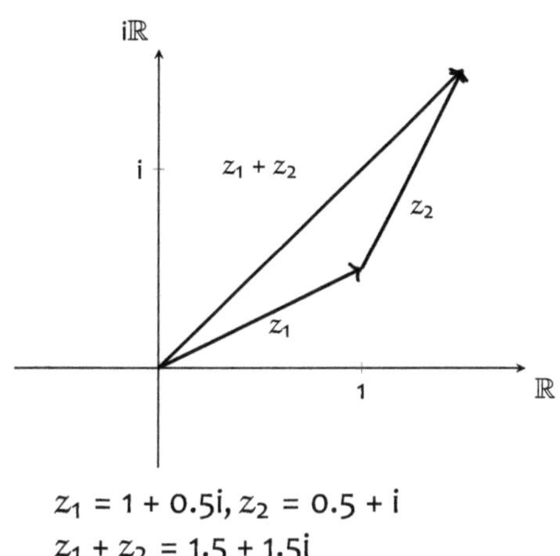


$$z_1 = 1 + 0.5i, z_2 = 0.5 + i$$
$$z_1 + z_2 = 1.5 + 1.5i$$

# How can you multiply complex numbers on the plane?

The distance of a complex number $z = a + bi$ from the origin is called the modulus of the complex number $|z| = |a + bi| = \sqrt{a^2 + b^2}$. It gives you the size of a complex number.

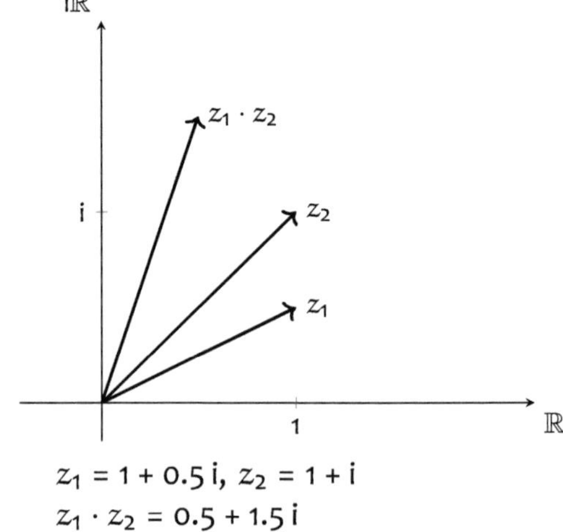



If you multiply two complex numbers, their moduli multiply, or, equivalently, the squares of the moduli multiply.

$$|(a + bi)(c + di)|^2$$
$$= |(ac - bd) + (ad + bc)i|^2$$
$$= (ac - bd)^2 + (ad + bc)^2$$
$$= a^2c^2 - 2abcd + bd^2 +$$
$$\quad + a^2d^2 + 2abcd + b^2c^2$$
$$= a^2c^2 + bd^2 + a^2d^2 + b^2c^2$$
$$= (a^2 + b^2)(c^2 + d^2)$$
$$= |a + bi|^2\, |c + di|^2$$

If you multiply two complex numbers, their moduli multiply and their angles add, as can be seen directly using Euler's formula and the polar representation $z = r\exp(i\varphi)$, but we won't touch on it.



$$z_1 = 1 + 0.5\,i,\quad z_2 = 1 + i$$
$$z_1 \cdot z_2 = 0.5 + 1.5\,i$$

# Why are complex numbers so important?

The remarkable thing is, that by adding one new number i and extending $\mathbb{R}$ to $\mathbb{C} = \{a + bi \mid a, b \in \mathbb{R}\} = \mathbb{R} + \mathbb{R}i$, all polynomial equations with real and/or complex coefficients can be solved within $\mathbb{C}$.

It can be shown that every equation of degree $n$ actually has exactly $n$ complex solutions, if counting multiple solutions according to their multiplicity.

The solutions of the equation $z^n = 1$ form the vertices of a regular $n$-gon, with one vertex at $z = 1$, i.e. on the positive real axis. They are called the $n$-th roots of unity.

Fundamental Theorem of Algebra:

Every equation
$$a_n x^n + a_{n-1} x^{n-1} + \ldots + a_0 = 0$$
has a solution $x \in \mathbb{C}$.

Quadratic equation:
$$x^2 + px + q = 0$$
$$x_{1,2} = -\frac{p}{2} \pm \sqrt{D}; \quad D = \left(\frac{p}{2}\right)^2 - q$$

$$D \begin{cases} > 0: & \text{2 real solutions} \\ = 0: & \text{1 double real solution} \\ < 0: & \text{2 complex solutions} \end{cases}$$

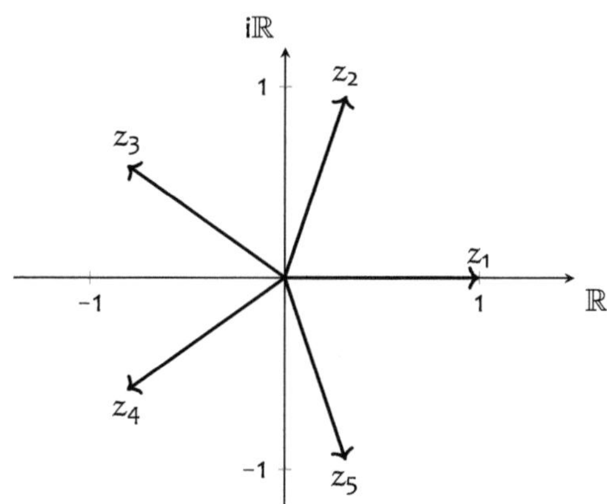


The 5 fifth roots of unity $z^5 = 1$:
$$z_1 = 1$$
$$z_2 = \cos\left(\tfrac{2\pi}{5}\right) + i\sin\left(\tfrac{2\pi}{5}\right)$$
$$z_3 = \cos\left(\tfrac{4\pi}{5}\right) + i\sin\left(\tfrac{4\pi}{5}\right)$$
$$z_4 = \cos\left(\tfrac{6\pi}{5}\right) + i\sin\left(\tfrac{6\pi}{5}\right)$$
$$z_5 = \cos\left(\tfrac{8\pi}{5}\right) + i\sin\left(\tfrac{8\pi}{5}\right)$$

# How can you represent functions by a power series?

As the derivative gives the linearization at a point of a differentiable function, the function value $f(x)$ can be written using the function value $f(0)$ and the derivative $f'(0)$.

$$f(x) = f(0) + f'(0)x + \text{remainder}$$

This can be refined by adding quadratic, cubic, etc. terms.

$$f(x) = f(0) + f'(0)x + \frac{f''(0)}{2}x^2 + \frac{f'''(0)}{6}x^3 + \ldots$$

The coefficients are chosen so that all derivatives at the point $x = 0$ are the same on both sides of the equation.

$$f'(x) = f'(0) + f''(0)x + \frac{f'''(0)}{2}x^2 + \ldots$$

is exact for $x = 0$.

$$f''(x) = f''(0) + f'''(0)x + \ldots$$

is exact for $x = 0$.

The corresponding series is called the Taylor series. For many relevant functions, this series equals the function. When it's the case, this is of course only valid for those values of $x$ for which the series converges.

$$f(x) = f(0) + f'(0)x + \frac{f''(0)}{2!}x^2 + \frac{f'''(0)}{3!}x^3 + \ldots + \frac{f^{(n)}(0)}{n!}x^n + \ldots$$

You can also expand a function into a Taylor series about a point other than $x = 0$. To do so, you take the value of the function and its derivatives at this point $a$, and what was previously $x$ is now called $h$.

$$f(a + h) = f(a) + f'(a)h + \frac{f''(a)}{2!}h^2 + \ldots + \frac{f^{(n)}(a)}{n!}h^n + \ldots$$

If you now set $a + h = x$, you obtain the Taylor series for $f(x)$ at the point $a$ (above we had $a = 0$).

$$f(x) = f(a) + f'(a)(x - a) + \frac{f''(a)}{2!}(x-a)^2 + \frac{f'''(a)}{3!}(x-a)^3 + \ldots + \frac{f^{(n)}(a)}{n!}(x - a)^n + \ldots$$

# How do you calculate Taylor series?

The Taylor series of the exponential function about the point $x = 0$ is obtained by calculating the derivatives at $x = 0$. We have already motivated this series earlier with the definition of $e$ and $e^x$.

$$f(x) = e^x, f'(x) = e^x, f''(x) = e^x, \ldots$$
$$f(0) = f'(0) = f''(0) = \ldots = 1$$

$$e^x = 1 + x + \frac{x^2}{2!} + \frac{x^3}{3!} + \frac{x^4}{4!} + \ldots$$

Similarly, you obtain the Taylor series of the sine function.

$$f(x) = \sin(x), f'(x) = \cos(x),$$
$$f''(x) = -\sin(x), f'''(x) = -\cos(x), \ldots$$
$$f(0) = 0, f'(0) = 1,$$
$$f''(0) = 0, f'''(0) = -1, \ldots$$

$$\sin(x) = x - \frac{x^3}{3!} + \frac{x^5}{5!} \pm \ldots$$

In the same way, you can also obtain the Taylor series of the cosine function.

$$f(x) = \cos(x), f'(x) = -\sin(x),$$
$$f''(x) = -\cos(x), f'''(x) = \sin(x), \ldots$$
$$f(0) = 1, f'(0) = 0,$$
$$f''(0) = -1, f'''(0) = 0, \ldots$$

$$\cos(x) = 1 - x^2 + \frac{x^4}{4!} - \frac{x^6}{6!} \pm \ldots$$

The logarithm is not defined for 0. Therefore, you expand it into a Taylor series about the point $x_0 = 1$. You can do this as above.

$$f(x) = \ln(x), f'(x) = x^{-1},$$
$$f''(x) = -x^{-2}, f'''(x) = 2x^{-3}, \ldots$$
$$f(1) = 0, f'(1) = 1, f''(1) = -1!,$$
$$f'''(1) = 2!, f''''(1) = -3!, \ldots$$

$$\ln(1 + x) = x - \frac{x^2}{2} + \frac{x^3}{3} - \frac{x^4}{4} \pm \ldots$$

A simpler derivation can be achieved by integrating the geometric series. The constant of integration is 0 here, as you can see by substituting $x = 0$.

$$\frac{1}{1 + x} = 1 - x + x^2 - x^3 \pm \ldots$$

$$\ln(1 + x) = \int \frac{1}{1 + x}\, dx$$

$$\ln(1 + x) = x - \frac{x^2}{2} + \frac{x^3}{3} - \frac{x^4}{4} \pm \ldots$$

With this, you can, e.g., calculate $\ln(2)$. However, there are faster methods.

$$\ln(2) = 1 - \frac{1}{2} + \frac{1}{3} - \frac{1}{4} \pm \ldots$$

# How is the Taylor series related to the exponential function of the differential operator?

If you look closely at the general formula for the Taylor series, you will notice that it looks very similar to the Taylor series of the function $e^x$.

$$f(a) = f(0) + f'(0)a + \frac{f''(0)}{2!}a^2 + \dots$$

$$\exp(x) = 1 + x + \frac{x^2}{2!} + \dots$$

Of course, , this is because all derivatives of the function $e^x$ at $x = 0$ are equal to 1.

$$f(x) = \exp(x):$$
$$f(0) = f'(0) = f''(0) = \dots = 1$$

But it also has a deeper meaning: you can write the general Taylor series as the exponential function of the differential operator, $\frac{d}{dx}$.

$$f(x + a) = \exp\left(a\frac{d}{dx}\right) f(x) =$$

$$= \left(1 + a\frac{d}{dx} + \left(a\frac{d}{dx}\right)^2 + \dots\right) f(x)$$

$$= f(x) + a\frac{d}{dx}f(x) + a^2 \frac{d^2}{dx^2} f(x) + \dots$$

For $x = 0$ this yields the Taylor series from above.

The change of the function for a very small shift, $h$, is $h\frac{d}{dx}f(x)$.

$$\exp\left(h\frac{d}{dx}\right) f(x) - f(x) =$$
$$\left(1 + h\frac{d}{dx}\right) f(x) - f(x) = h\frac{d}{dx}f$$

On its own, without $f(x)$, the operator is $h\frac{d}{dx}$. We say that $\frac{d}{dx}$ is the infinitesimal generator of translations.

The operator for a finite translation $(a = nh = n\frac{a}{n})$ is obtained by exponentiation, which in the limit $n \rightarrow \infty$, gives the exponential function.

$$\lim_{n\to\infty}\left(1 + \frac{a}{n}\frac{d}{dx}\right)^n = \exp\left(a\frac{d}{dx}\right)$$

This can be generalized and plays a major role in many areas of physics under the keyword "Lie group".

# How is $e^{ix}$ related to $\sin(x)$ and $\cos(x)$?

If you calculate $e^{ix}$ by substituting $ix$ into the Taylor series of $e^x$, you can split the series into a real part and an imaginary part.

$$e^{ix} = 1 + ix + \frac{(ix)^2}{2!} + \frac{(ix)^3}{3!} + \frac{(ix)^4}{4!} + \ldots$$

$$= 1 + ix - \frac{x^2}{2!} - i\frac{x^3}{3!} + \frac{x^4}{4!} + \ldots$$

$$= 1 - \frac{x^2}{2!} + \frac{x^4}{4!} \pm \ldots + ix - i\frac{x^3}{3!} \pm \ldots$$

You can see that the real part is exactly the Taylor series of the cosine and the imaginary part, that of the sine.

$$e^{ix} =$$

$$= \underbrace{1 - \frac{x^2}{2!} + \frac{x^4}{4!} \pm \ldots}_{\cos(x)} + i\underbrace{\left( x - \frac{x^3}{3!} \pm \ldots \right)}_{\sin(x)}$$

This yields Euler's formula.

$$e^{ix} = \cos(x) + i\sin(x)$$

For the special case of $x = \pi$, we obtain an equation that has been selected in various polls as the most beautiful formula in the world. It connects the two most important mathematical constants, e and $\pi$, with zero, one, and the imaginary unit, i.

$$\boxed{e^{i\pi} + 1 = 0}$$

$$e = \lim_{n \to \infty} \left( 1 + \frac{1}{n} \right)^n$$

$$i = \sqrt{-1}$$

$\pi$ = circumference / diameter

1: neutral element of multiplication

0: neutral element of addition

Euler's formula forms a bridge between analysis, algebra, and geometry, and is also a key cornerstone in the proof that $\pi$ is transcendental, i.e. not a root of any polynomial with integer coefficients.

$$e^{i\pi} + 1 = 0$$

e : analysis

i : algebra

$\pi$ : geometry

By the way, there is still a lot we do not know about the relationship between $\pi$ and e.

Open question:

Is $\pi + e$ a rational number?

# How can you understand $e^{i\pi} + 1 = 0$ geometrically?

We want to explain the equation in a visual-geometric way. This explanation can be expanded into a proof.

$$e^{i\pi} + 1 = 0$$

First, we use the formula for $e^x$, which is based on the definition of e, and set $x = i\pi$.

$$e^x = \lim_{n\to\infty} \left(1 + \frac{x}{n}\right)^n$$

$$e^{i\pi} = \lim_{n\to\infty} \left(1 + \frac{i\pi}{n}\right)^n$$

The complex number in the parentheses is a scaling-rotation: a rotation by an angle of approximately $\pi/n$ (for large $n$, $\tan(\pi/n) \approx \pi/n$) and a scaling by the factor $\sqrt{1 + (\pi/n)^2}$.

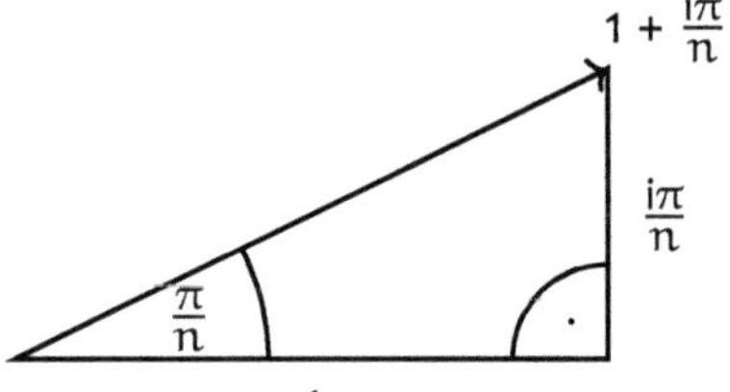

To obtain $(1 + i\pi/n)^n$, one performs $n$ times such a scaling-rotation (on the right, $n = 10$), starting with the triangle at the bottom, right end of the spiral. In total, this results in a rotation by approximately $\pi = 180°$ and a scaling by $(\sqrt{1 + (\pi/n)^2})^n$.

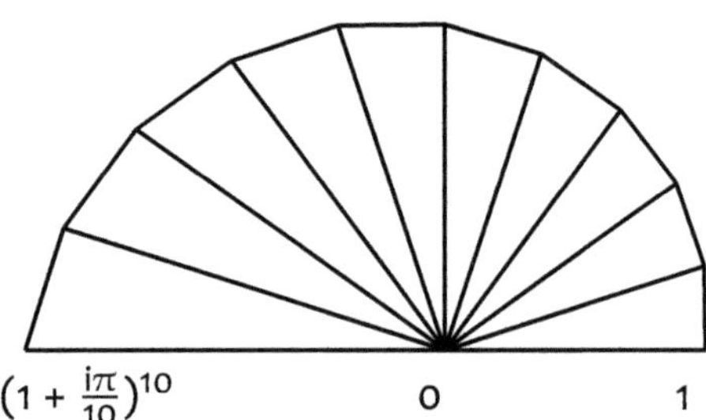

It can be shown that, as $n \to \infty$, we get a rotation of exactly $\pi = 180°$, which is a multiplication by $-1$.

The deviation of the scaling factor from 1 is only proportional to $1/n^2$. For $n$ scalings and $n \to \infty$, the total scaling factor becomes 1. The rotation angle for large $n$ is $\approx \pi/n$. For $n$ rotations and $n \to \infty$, the total angle becomes $\pi$.

$$e^{i\pi} = \lim_{n\to\infty} \left(1 + \frac{i\pi}{n}\right)^n = -1 \quad \text{or}$$

$$e^{i\pi} + 1 = 0$$

Thus, the desired formula is obtained.

# What is a differential equation?

An ordinary differential equation (ODE) is an equation that contains an initially unknown function and its derivative(s).

$$y''(x) + y(x) = 0$$
differential equation for oscillations

$$\left((y'')^{-2/3}\right)''' = 0$$
differential equation for conic sections

The solution of an ODE is a function which, when substituted along with its derivatives into the ODE, satisfies the equation.

$y(x) = \sin(x)$ satisfies the ODE $y'' + y = 0$, because
$y' = \cos(x)$, $y'' = -\sin(x) = -y$.

The highest derivative that occurs is called the order of a differential equation.

$$y''(x) + y(x) = 0$$
is a second-order ODE.

A simple class of differential equations are those that are linear in the function and its derivatives.

$$y''(x) + y(x) = 0: \text{ linear ODE}$$

$$\left((y'')^{-2/3}\right)''' = 0: \text{ nonlinear ODE}$$

One can show that the solutions of linear differential equations form a vector space.

If $y_1(x)$ and $y_2(x)$ are solutions of $y''(x) + y(x) = 0$, then so are $ay_1(x)$, $a \in \mathbb{R}$ and $y_1(x) + y_2(x)$. Therefore, the solutions form a vector space.

The dimension of the space of solutions is equal to the order of the differential equation. Basis vectors of the vector space of solutions are called fundamental solutions.

The solutions of $y'' + y = 0$ form a 2-dimensional vector space, which is spanned by the fundamental solutions, $\sin(x)$ and $\cos(x)$.

The general solution of an $n^{\text{th}}$ order differential equation contains $n$ parameters.

The general solution of
$y'' + y = 0$ is
$y(x) = a\sin(x) + b\cos(x), \ a, b \in \mathbb{R}$.

# What is meant by "configuration space"? (I)

A geometric configuration often allows for degrees of freedom, i.e. parameters that can be varied independently.

We specify each of these configurations by one point in a configuration space. All configurations form a configuration space, where small changes in the configuration correspond to neighboring points in the configuration space.

The configuration space of all congruent triangles is 3-dimensional, that of all quadrilaterals is 5-dimensional.

If you also abstract from the size, then for triangles two degrees of freedom remain. The space of all triangle shapes is thus 2-dimensional.

In general, configuration spaces aren't vector spaces, i.e. they aren't linear spaces. Often they are manifolds.

Since they locally look like an $\mathbb{R}^n$, you can assign a dimension to them ($n$ = number of DOF). However, manifolds of the same dimension can (in contrast to vector spaces) look completely different globally.

The configuration of two points on the $x$-axis has two degrees of freedom (DOF), since you can change the two $x$-values independently of each other.

If the points are indistinguishable, the configuration is uniquely determined by the pair $(x_{min}, x_{max}) \in \mathbb{R}^2$, where $x_{max} \geq x_{min}$. The configuration space is a half-plane.

Three points in $\mathbb{R}^2$ have $3 \cdot 2 = 6$ DOF If you abstract from position, i.e. translation (2 DOF) and orientation (1 DOF), there remain $6 - 2 - 1 = 3$ DOF for the shape and size of the triangle.

The size is one DOF, thus:
$$3 \cdot 2 - \underbrace{3}_{\text{position}} - \underbrace{1}_{\text{size}} = 2 \text{ DOF}$$

The shape of a triangle can, for example, be described by two angles.

Manifolds look like a vector space locally (in the small), but look different globally (in the large). If you, as an ant, walk on a beach ball, it feels like a plane to you.

A spatial pendulum and a planar double pendulum both have a 2-dimensional configuration space, but the first is a sphere, the second a torus (donut).

# What is meant by "configuration space"? (II)

Configuration spaces can also describe the set of all positions of a fixed geometric object in space.

The number of degrees of freedom (DOF) is the number of independent quantities you need in order to specify the position including orientation of the geometric object.

A line in the plane has 2 degrees of freedom, for example the two intercepts with the $x$- and $y$-axis.

A plane in space has 3 degrees of freedom, for example the three intercepts with the axes.

A line in space has 4 degrees of freedom. First, consider a line through the origin in space; it has 2 degrees of freedom (two angles, i.e. a point on the 2-dimensional sphere). To reach an arbitrary line, you can additionally shift the line through the origin in the 2-dimensional plane perpendicular to its direction.

So in total there are 4 degrees of freedom.

A dumbbell in the plane, i.e. two points ($2 \cdot 2$ DOF) connected by a fixed rod (1 constraint), has $2 \cdot 2 - 1 = 3$ DOF

To specify the position of the dumbbell, you can give its midpoint (2 parameters) as well as its angle to the $x$-axis (1 parameter), so in total you need 3 parameters.

Alternatively: 2 points in $\mathbb{R}^2$ have $2 \cdot 2$ DOF, but you can shift any point along the line, i.e. you lose $2 \cdot 1$ DOF, so you get again $2 \cdot 2 - 2 \cdot 1 = 2$ DOF

Alternatively: 3 points in $\mathbb{R}^3$ have $3 \cdot 3$ DOF, but you can shift any point in the plane, i.e. you lose $3 \cdot 2$ DOF, so get again $3 \cdot 3 - 3 \cdot 2 = 3$ DOF

Alternatively: 2 points in $\mathbb{R}^3$ have $2 \cdot 3$ DOF, but you can shift any point along the line, i.e. you lose $2 \cdot 1$ DOF, so you get again $2 \cdot 3 - 2 \cdot 1 = 4$ DOF

If you are interested in this in more detail, look up the keyword "Plücker coordinates" on the internet.

# What is geometric algebra?

In vector algebra we can add and subtract vectors and multiply them by scalars. We can also multiply vectors by each other, but dividing them by each other is not possible.

$$\mathbf{a} + \mathbf{b} \in V$$
$$r\mathbf{a} \in V$$
$$\mathbf{a} \cdot \mathbf{b} \in \mathbb{R}$$
$$\mathbf{a} \times \mathbf{b} \in V$$
$$\mathbf{a} \cdot \mathbf{x} = c \;\not\Longrightarrow\; \text{unique } \mathbf{x}$$
$$\mathbf{a} \times \mathbf{x} = \mathbf{c} \;\not\Longrightarrow\; \text{unique } \mathbf{x}$$

A real number has a "size" (the absolute value), no position in space (no direction), but an orientation.

The orientation of a number is its sign.

A vector also has a "size" (the length), has a 1-dimensional direction (the line through the origin it defines), and has an orientation.

$\mathbf{a}$ and $-\mathbf{a}$ define the same line, but have different orientations.

Now, besides 0-dimensional and 1-dimensional objects, there are also 2- and higher-dimensional objects.

A parallelogram spanned by two vectors also has a "size" (the area), a 2-dimensional position in space (the "direction" of the plane), and an orientation.

If these geometric objects exist in all dimensions, it would be useful to also have corresponding algebraic objects that we can calculate with. Geometric algebra makes this possible. Classical vector algebra is only the 1-dimensional part of it.

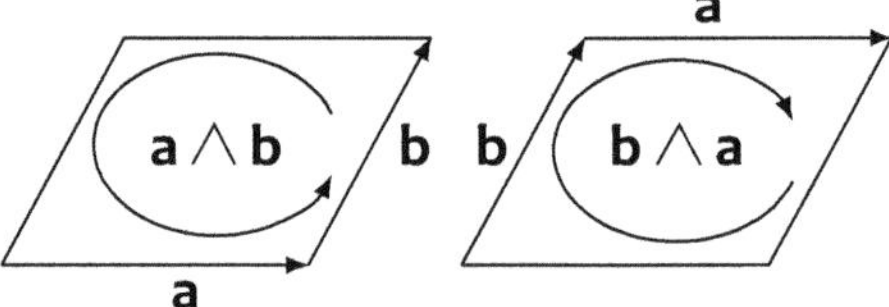


The orientation of a parallelogram spanned by $\mathbf{a}$ and $\mathbf{b}$, $\mathbf{a} \wedge \mathbf{b}$, depends on the order of the vectors.

To generate these higher-dimensional geometric objects, we will define a "geometric product" on the following pages.

As a byproduct, it will also become possible to divide by vectors.

If you want to learn more, look up the keyword "geometric algebra" on the internet.

# What is the geometric product?

We define a geometric product of two vectors. It should be associative and distributive over addition. In general, it is not commutative.

**ab**: geometric product
$$\mathbf{a}(r\mathbf{b} + s\mathbf{c}) = r\mathbf{ab} + s\mathbf{ac}$$
$$\mathbf{ab} \neq \mathbf{ba}$$

We decompose the geometric product into a symmetric and an antisymmetric part.

$$\mathbf{ab} = \tfrac{1}{2}(\mathbf{ab} + \mathbf{ba}) + \tfrac{1}{2}(\mathbf{ab} - \mathbf{ba})$$

The first term is symmetric and bilinear like the scalar product, so we set it equal to the scalar product, which is also called the inner product.

$$\tfrac{1}{2}(\mathbf{ab} + \mathbf{ba}) = \mathbf{a} \cdot \mathbf{b}$$
The inner product is a scalar.

You can see that the geometric product of a vector with itself yields a non-negative number, the square of its length.

$$\mathbf{aa} = \mathbf{a} \cdot \mathbf{a} + 0 = |\mathbf{a}|^2$$

We refer to the antisymmetric part as the exterior product, and we denote it with a wedge symbol. I will explain its meaning on the next page.

$$\tfrac{1}{2}(\mathbf{ab} - \mathbf{ba}) = \mathbf{a} \wedge \mathbf{b}$$
The exterior product of two vectors is also called bivector.

We can now express the geometric product of two vectors as the sum of the inner product and the exterior product.

$$\mathbf{ab} = \underbrace{\mathbf{a} \cdot \mathbf{b}}_{\text{scalar}} + \underbrace{\mathbf{a} \wedge \mathbf{b}}_{\text{bivector}}$$

It's not a problem to add these completely different quantities; you just can't compare them.

In the case of a complex number, real and imaginary numbers are also added.

# What does the exterior product mean?

We start from the definition of the geometric product as the sum of the inner product and the exterior product. Additionally, we take the same equation with the two vectors swapped.

$$\mathbf{ab} = \mathbf{a} \cdot \mathbf{b} + \mathbf{a} \wedge \mathbf{b}$$
$$\mathbf{ba} = \mathbf{b} \cdot \mathbf{a} + \mathbf{b} \wedge \mathbf{a} = \mathbf{a} \cdot \mathbf{b} - \mathbf{a} \wedge \mathbf{b}$$

If you multiply the last two equations, you can apply the third binomial formula on the right side of the equation. Then you can simplify the left side and write the scalar product on the right using lengths and angles.

$$\mathbf{abba} = (\mathbf{a} \cdot \mathbf{b})^2 - (\mathbf{a} \wedge \mathbf{b})^2$$
$$\mathbf{a}|\mathbf{b}|^2\mathbf{a} = (\mathbf{a} \cdot \mathbf{b})^2 - (\mathbf{a} \wedge \mathbf{b})^2$$
$$|\mathbf{a}|^2|\mathbf{b}|^2 = |\mathbf{a}|^2|\mathbf{b}|^2 \cos^2(\varphi) - (\mathbf{a} \wedge \mathbf{b})^2$$

The square of the exterior product is thus the negative square of the area of the parallelogram spanned by **a** and **b**.

$$(\mathbf{a} \wedge \mathbf{b})^2 = -|\mathbf{a}|^2|\mathbf{b}|^2 \sin^2(\varphi)$$

The magnitude of the exterior product $\mathbf{a} \wedge \mathbf{b}$ is therefore the area.

$$|\mathbf{a} \wedge \mathbf{b}| = |\mathbf{a}|\,|\mathbf{b}|\,|\sin(\varphi)|$$

In three dimensions, the exterior product can be related to the cross product.

$$\mathbf{a} \wedge \mathbf{b} = i\,\mathbf{a} \times \mathbf{b}$$

This works in three dimensions, but not in four or more dimensions. In two dimensions, the cross product cannot really be defined, since there is no third dimension.

This works in $\mathbb{R}^3$, since in this case two vectors define, up to orientation, a unique direction perpendicular to both.

Therefore, the exterior product is much more general and systematic than the cross product.

If you want to learn more, look up "exterior product" on the internet.

# How do you concretely calculate the geometric product?

Since the geometric product is linear, we can calculate it for two vectors if we know how to do it for the basis vectors.

The geometric products of two basis vectors can be calculated directly from the definition.

Thus, you can calculate the geometric product of any two vectors using the linearity.

There are only four basis vectors, since higher basis vector products can be reduced to these.

In 3 dimensions, this works similarly, you just get more terms.

The set $\{a+b\mathbf{e}_x+c\mathbf{e}_y+d\mathbf{e}_x\mathbf{e}_y, a, b, c, d \in \mathbb{R}\}$ forms an algebra under addition and the geometric product.

Let us calculate the square of the bivector $\mathbf{e}_x\mathbf{e}_y$. This results in $-1$.

The even Clifford subalgebra $\{a + b\mathbf{e}_x\mathbf{e}_y, a, b \in \mathbb{R}\}$ in 2 dimensions is therefore isomorphic to the complex numbers.

In 3 dimensions, something new arises.

In two dimensions, $\mathbf{ab} =$
$$= (a_x\mathbf{e}_x + a_y\mathbf{e}_y)(b_x\mathbf{e}_x + b_y\mathbf{e}_y) =$$
$$= a_xb_x\mathbf{e}_x\mathbf{e}_x + a_yb_x\mathbf{e}_x\mathbf{e}_y +$$
$$+ a_yb_x\mathbf{e}_y\mathbf{e}_x + a_yb_y\mathbf{e}_y\mathbf{e}_y$$

$$\mathbf{e}_x\mathbf{e}_x = \underbrace{\mathbf{e}_x \cdot \mathbf{e}_x}_{=1} + \underbrace{\mathbf{e}_x \times \mathbf{e}_x}_{=0} = 1$$
$$\mathbf{e}_x\mathbf{e}_y = \underbrace{\mathbf{e}_x \cdot \mathbf{e}_y}_{=0} + \underbrace{\mathbf{e}_x \times \mathbf{e}_y}_{\text{bivector}} = -\mathbf{e}_y\mathbf{e}_x$$

$$(\mathbf{e}_x + 2\mathbf{e}_y)(3\mathbf{e}_x + 4\mathbf{e}_y) =$$
$$= 3\mathbf{e}_x\mathbf{e}_x + 4\mathbf{e}_x\mathbf{e}_y + 6\mathbf{e}_y\mathbf{e}_x + 8\mathbf{e}_y\mathbf{e}_y$$
$$= 3 + 4\mathbf{e}_x\mathbf{e}_y - 6\mathbf{e}_x\mathbf{e}_y + 8$$
$$= 11 - 2\mathbf{e}_x\mathbf{e}_y$$

$$\{1, \mathbf{e}_x, \mathbf{e}_y, \mathbf{e}_x\mathbf{e}_y\}$$

In any product of two vectors, we only find the "even" basis vectors $1$ and $\mathbf{e}_x\mathbf{e}_y$.

A basis in 3 dimensions is:
$$\{1, \mathbf{e}_x, \mathbf{e}_y, \mathbf{e}_z, \mathbf{e}_x\mathbf{e}_y, \mathbf{e}_x\mathbf{e}_z, \mathbf{e}_y\mathbf{e}_z, \mathbf{e}_x\mathbf{e}_y\mathbf{e}_z\}$$

This is called the 2-dimensional Clifford algebra, and it is also defined in higher dimensions.

$$(\mathbf{e}_x\mathbf{e}_y)^2 = (\mathbf{e}_x\mathbf{e}_y)(\mathbf{e}_x\mathbf{e}_y) =$$
$$= -(\mathbf{e}_x\mathbf{e}_y)(\mathbf{e}_y\mathbf{e}_x)$$
$$= -\mathbf{e}_x\underbrace{\mathbf{e}_y\mathbf{e}_y}_{=1}\mathbf{e}_x = -\underbrace{\mathbf{e}_x\mathbf{e}_x}_{=1} = -1$$

$$\{a + b\mathbf{e}_x\mathbf{e}_y\} \cong \{a + bi\} \cong \mathbb{C}$$

keywords: quaternions, Pauli matrices

# Table of Overviews


Addition and multiplication....................................................12
Addition and subtraction ....................................................5
Antiderivatives...............................................................193
Arithmetic of fractions ......................................................21
Curve sketching ..............................................................175
Differentiation rules ........................................................163
Exponents, roots, logarithms ................................................38
Extrema and inflection points ...............................................170
Graph of function and derivative.............................................171
Injective, surjective, and bijective.........................................122
Multiplication and division ..................................................20
Parametric and implicit equations ...........................................214
Products of vectors in different notations...................................254
Relative positions of points, lines, and planes ............................253



© The Editor(s) (if applicable) and The Author(s), under exclusive license
to Springer-Verlag GmbH, DE, part of Springer Nature 2026
A. Gründers, *Math Made Clear: From the Basics to Calculus*,
https://doi.org/10.1007/978-3-662-73221-2

# If you are searching for terms or symbols, look here!


**Symbols**

| | |
|---|---|
| $\in$ | 2 |
| $\notin$ | 2 |
| $\emptyset$ | 44 |
| $=$ | 2 |
| $\neq$ | 7 |
| $+$ | 3 |
| $-$ | 4 |
| $\cdot$ | 6 |
| $:$ | 13 |
| $/$ | 13 |
| $\div$ | 13 |
| $!$ | 126 |
| $\propto$ | 89 |
| $\approx$ | 89 |
| $\cong$ | 65, 89 |
| $(\dots)$ | 8 |
| $\{\dots\}$ | 2 |
| $(\dots)$ vs. $\{\dots\}$ | 45 |
| $(a, b)$ (interval) | 46 |
| $[a, b]$ (interval) | 46 |
| $(a, b]$ (interval) | 46 |
| $[a, b)$ (interval) | 46 |
| $\wedge$ | 40 |
| $\vee$ | 40 |
| $^{-}$ (e.g. $\overline{A}$) | 256 |
| $^{-}$ (e.g., $\overline{A}$) | 40 |
| $\cap$ | 45 |
| $\cup$ | 45 |
| $\setminus$ | 45 |

| | |
|---|---|
| $\Rightarrow$ | 41 |
| $\Leftarrow$ | 41 |
| $\Leftrightarrow$ | 43 |
| $'$ (e.g. $f'$) | 146 |
| $\sum$ | 125 |
| $\int$ | 179 |
| $\infty$ | 46, 132 |
| $\forall$ | 275 |
| $\exists$ | 275 |
| $2 \times 2$ determinant | 227 |
| $3 \times 3$ determinant | 229 |

**A**

| | |
|---|---|
| abelian group | 216 |
| absolute value | 32 |
| absolute value equation | 59 |
| adding, subtracting fractions | 17 |
| addition | 3, 5 |
| addition of matrices | 220 |
| addition of random variables | 262 |
| addition of vectors | 200 |
| adjacent side | 68 |
| algebra | 276 |
| algebraic structures | 276 |
| analytic Geometry | 196 |
| and | 40 |
| angle between vectors | 205 |
| angle measurement | 79 |
| angles in a triangle | 67 |
| antiderivative | 181, 184, 193 |



© The Editor(s) (if applicable) and The Author(s), under exclusive license
to Springer-Verlag GmbH, DE, part of Springer Nature 2026
A. Gründers, *Math Made Clear: From the Basics to Calculus*,
https://doi.org/10.1007/978-3-662-73221-2

applications of mathematics  84
arcsin  111
arctan  162
area  74
area calculation  178, 185
area of a circle  80, 189, 190
area of a parallelogram  231
area of a triangle  74, 231
argument of a function  96
arithmetic mean  168
arithmetic of fractions  21
asymptote  119
axial symmetry  107

**B**
base  28
base value  90
base-2 logarithm  34
basis of a vector space  217
Bayes' theorem  258
Bernoulli distribution  262
bijective  99, 122
binomial coefficients  126
binomial distribution  267
binomial distribution as a sum  262
binomial formulas  30, 75, 126
binomial theorem  126
bivector  291
bounded sequence  134
bra-ket notation  206

**C**
calculating with units of meas.  92
capital  91
cardinality  44, 274
cartesian coordinate system  81
cartesian product  45
case distinction  59
Cayley-Menger determinant  231

central limit theorem  270
chain rule  159
chicken problem  86
circle, area  80
circle, circumference  78
circle, parametrization  214
circumference of a circle  78
circumference of a segment  79
Clifford algebra  293
closed interval  46
column vector  197
common logarithm  33
complement of a set  256
complementary event  256
complete orthonormal system  206
completeness of $\mathbb{R}$  134
completing the square  56
complex numbers  277, 278
complex plane  279
composite function  158
compound interest  91
concave down  166
concave up  166
conditional probability  258
cone  76
configuration space  288, 289
continuous  140, 141, 144
continuous distribution  268
continuous extension  142
continuum problem  274
convergence  129–131, 133
convergence criteria  137
convergence rules  133
conversion of line equ.  212, 213
conversion of plane equ.  212, 213
conversion to radians  79
coordinate system  81
cos  71

cosine 68, 71, 72
countably infinite 274
criteria for continuity 143
cross product 232, 233, 254
cubic equations 57
cubic meter 92
curly bracket 2
curvature 165, 166
curve sketching 174, 175
cylinder 76

**D**

data science 210
decimal fractions 22
decimal numbers 22
definite integral 179
degrees of freedom 288
denominator 15
density function 268
dependent variable 96
derivative 146, 147
derivative as a linear map 218
derivative as linearization 154
derivative of inverse funct. 161, 162
deriving physical formulas 93
determinant 61, 226, 227, 229
determinant, calculation 230
determinant, properties 228
differentiable 148
differential 155
differential equation 287
differentiation rules 163
dimension of a vector space 217
dimension of the solution 245
dimension theorem 236, 237, 246
dimensional analysis 93
dimensions 77
direct proof 47

discrete random variable 260, 261
distance between two points 81
distance line to line 252
distance line to plane 252
distance plane to plane 252
distance point to line in 2D 250
distance point to line in 3D 251
distance point to plane 250
distr. of a random variable 260
distribution 261
distribution function 268
distributive law 10
divergence 132
divided by 13
divisible 15
division 13, 20
division by zero 14
domain 96
dot product 204, 254
dot product and projection 206
double zero 116
doubling of capital 91
$dx$ 155
dyadic product 223, 254

**E**

$\in$ 2
$e$ 151, 152, 285
$\epsilon$-neighborhood 32, 129
element 2
endomorphism 237
epsilon-neighborhood 32, 129
equations of the fifth degree 276
equivalence rearrangements 50–52
Euler's formula 285
Euler's number 151, 152, 285
event 256
$\exp(x)$ 160

expanding 12, 25
expanding fractions 16
expected value 263
exponent 28, 38
exponential equation 58
exponential function 120
exterior product 292
extrema 164, 170, 171, 175

**F**

factorial 126
factoring 12, 25
false 40
field 216, 276
for all 275
Fourier decomposition 206
fraction arithmetic 17–19, 21
fractional numbers 15
fractions 15, 16
function 96
function, representation 97
fundamental theorem 182, 183

**G**

Galois theory 276
Gaussian bell curve 269
Gaussian distribution 269–271
geometric algebra 290, 293
geometric mean 168
geometric product 291
geometric series 136
graph of a function 97
group 216, 276
guessing solutions 57

**H**

harmonic series 135
Heron's formula 231
Hesse normal form 209, 210

Hesse normal form of a plane 211
higher-degree equations 57
HNF of a line 209, 210
HNF of a plane 211
hyperplane 210
hypotenuse 66, 68

**I**

i 277
identity matrix 222, 224
if and only if 43
image of a linear map 234
imaginary unit 277
implication 41
implicit equations 214
improperly integrable 192
indefinite integral 180
independent events 257
independent variable 96
indices 124
indirect proof 47
inequalities, important 169
inequality 60
infinite 274
infinite series 135
infinite-dimensional 217, 237
infinity 46, 132
inflection points 167, 170, 171, 175
injective 98, 99, 122
integers 4
integrable 191
integral table 193
integration 181
integration by parts 186, 187
intercept form 104
intercept theorem 65, 67
interest 90, 91
intersection line 249

intersection of line and plane   248
intersection of two lines   247
intersection of two planes   249
intersection with   45
interval   46
inverse function   111
inverse matrix   222, 224, 225
inverse of a matrix   243
inverse of exponentiation   37
invertible matrix   224
irrational numbers   23

**J**

jump   142

**K**

kernel of a linear map   234
kilometers per hour   92

**L**

Laplace expansion theorem   230
Laplace experiment   259
law of cosines   73
law of large numbers   263
law of sines   73
laws of exponents   28
lcm   18
LEAN   275
least common multiple   18
left shift   237
leg   66, 68
length of a vector   198
Lie group   284
lim   131, 138
limit   131
limit of a function   138
limit of a sequence   131
limit rules   139
limit theorems   133, 139

line and line   252, 253
line and plane   252, 253
line equation in 3D   251
line in the plane   103
line, HNF of   209, 210
line, parametric form   207
linear algebra   196
linear approximability   148
linear combination   201
linear equation   53, 54
linear factors   114
linear function   100, 102
linear hull   201
linear map   196, 218
linear map, image   234
linear map, kernel   234
linear space   215
linear system   238
linear system of equations   240
linear system, gen. solution   242
linear system, part. solution   242
linearly dependent   202
linearly independent   203
log   33
logarithm   33, 38
logarithm laws   36
logarithmic function   121
logic   40

**M**

manifold   288
mathematical induction   128
mathematics and reality   84
matrix   61, 215, 219
matrix equation   241
matrix multiplication   221–223, 225
maximum   164
mean   263

minimum 164
minimum or maximum? 165
minus 4
minus times minus 10, 11
modeling 85
modulus of a complex number 280
most beautiful formula 285
multipl. matrix and number 220
multipl. matrix with vector 219, 223
multipl. vector and number 199
multiplication 6, 20
multiplication before addition 7
multiplication of matrices 221, 222
multiplying 6
multiplying by zero 51
multiplying, dividing fractions 19

**N**
$\mathbb{N}$ 2
natural logarithm 34
natural numbers 2
necessary 42
negation 40
negative of a vector 200
normal distribution 269
normal vector 209
normal vector, normalized 209, 210
not 40
notations of vector products 254
number line 22
number of favorable cases 259
numerator 15

**O**
one, representation of 206
one-to-one correspondence 99
open interval 46
opposite side 68
optimization problems 172

or 40
order of operations 7, 8
oriented area 226
oriented volume 226
oscillation 287
outcome 256
outer product 291

**P**
$\pi$ 78, 80, 285
p-q-formula 56
$P_B(A)$ 258
parabola 105, 106
parallel to the x-axis 104
parallel to the y-axis 104
parallelogram, area 231, 232
parametric equations 214
parametric form of a line 207
parametric form of a plane 208
parametrization of the circle 214
parentheses 7–10, 12
parentheses for functions 35
Pauli matrices 293
per mille 90
percent 90
percentage rate 90
percentage value 90
perceptron 210
period of a pendulum 93
perpendicular lines 112
physical quantities 92
physics 84
piecewise defined function 144
placeholder 24
plane and plane 252, 253
plane, Hesse normal form of 211
plane, parametric form 208
plus 3

Plücker coordinates 289
Plücker form of line equation 251
point and line 250, 251, 253
point and plane 250, 253
point and point 253
point symmetry 107
polynomial 114
polynomial division 118, 119
polynomial functions 117
position vector 197, 198
power 28
power function 113
prime factors 15
prime number 15
prism 76
probability 257
probability density 268
probability density function 268
probability theory 255
product rule 156
product rule for indep. events 257
product set 45
projection and dot product 206
proof 47
proof assistant 275
proof by contradiction 47
proportional 89
Pythagoras 66
Pythagorean theorem 66, 73

**Q**

$\mathbb{Q}$ 15
quadratic equation 53, 55, 56
quadratic formula 56
quadratic function 106
quantifiers 275
quaternions 293
quotient rule 157

**R**

radian 79
radical equation 53
random experiment 256
random variable 260
random variables, addition 262
range 96
rank 235, 244
rank and linear system 245
rank-nullity theorem 236, 237, 246
rate of change 147
ratio test 137
rational function 117, 119
rational numbers 15, 23
rational parametrization 214
rearranging equations 26, 51, 52
reciprocal 19
recursive sequence 127
reducing fractions 16
reflection of functions 107, 111
right shift 237
right triangle 66–68
ring 276
rk 235
root 31, 38
root test 137
roots of unity 281
row echelon form 241, 244
row vector 197
rule of three 89

**S**

$\sum$ 125
$\int$ 179
saddle point 164, 167
Sarrus rule 230
scalar 216
scalar multiplication 216

scalar product                                      223
scalar triple product                               233
second degree equations                              53
second derivative                                   166
sensitivity                                          258
sequence                                            127
series                                              135
series representation                               282
set                                                2, 44
set difference                                       45
shifting functions                                  108
side lengths of a triangle                       68, 73
similar                                              65
sin                                              69, 70
sine                                          68, 69, 72
sine curve                                           72
sine values                                          70
sinh                                                 58
skew lines                                          247
slope of a curve                                    146
slope of a line                                     101
slope of the exp. function                     151, 152
slope triangle                                      101
slope-intercept form                                103
sol. of a lin. system               238–240, 242
solving a linear equation                            54
solving a quadratic equation                         55
solving word problems                            85, 86
span                                            201, 217
spanned space                                       201
specificity                                         258
sphere                                               76
square meter                                         92
square root                                          31
standard parabola                                   105
statement                                            40
statement form                                       41
stretching of functions                        109, 110

structure of lin. system                            242
substitution rule                              188, 189
subtraction                                        4, 5
subtraction of vectors                              200
sufficient                                           42
sum of angles in a triangle                          64
summation sign                                      125
surjective                                  98, 99, 122
system of lin. equations            238–241
system of linear equations                           61

**T**

tan                                                  71
tangent                            68, 71, 146, 147
Taylor series                              282–284
term                                                  5
test for disease                                    258
testing if 3 points are collinear   231
testing if 4 points are coplanar    231
there exists                                        275
transposed matrix                                   225
transposing                                         197
triangle's area                                 74, 231
trigonom. Pythagor. theorem             71
trigonometric parametrization  214
trivial linear combination                          203
true                                                 40
type of an equation                                  53

**U**

uncountably infinite                                274
union with                                           45
unit square                                          74
units                                                93
units of measures                                    92
unknown                                              24

**V**

variable                                          5, 24


304

vector     197, 198
vector product     232, 233
vector product, area     232
vector product, determinant     233
vector space     196, 215, 216
vector subspace     234
vector, addition     200
vector, multiplication     199
velocity     147
vertex     105, 106

volume     76
volume of a tetrahedron     231

**W**
whole numbers     4
word problems     85

**Z**
$\mathbb{Z}$     4
zero, multiple     116
zeros of polynomials     115